Quality Infrastructure of
Graphene
and Related Two-Dimensional Materials
Metrology

Quality Infrastructure of
Graphene
and Related Two-Dimensional Materials

Metrology

Editor

Lingling Ren
National Institute of Metrology, China

Translators

Lingling Ren
National Institute of Metrology, China
Xin Li
Shanghai International Studies University, China

ECUST PRESS

World Scientific

Published by

World Scientific Publishing Co. Pte. Ltd.

5 Toh Tuck Link, Singapore 596224

USA office: 27 Warren Street, Suite 401-402, Hackensack, NJ 07601

UK office: 57 Shelton Street, Covent Garden, London WC2H 9HE

Library of Congress Control Number: 2024024204

British Library Cataloguing-in-Publication Data
A catalogue record for this book is available from the British Library.

石墨烯材料质量技术基础: 计量
Originally published in Chinese by East China University of Science and Technology Press
Copyright © East China University of Science and Technology Press. 2021

QUALITY INFRASTRUCTURE OF GRAPHENE AND RELATED
TWO-DIMENSIONAL MATERIALS
Metrology

ISBN 978-981-12-9570-6 (hardcover)
ISBN 978-981-12-9571-3 (ebook for institutions)
ISBN 978-981-12-9572-0 (ebook for individuals)

For any available supplementary material, please visit
https://www.worldscientific.com/worldscibooks/10.1142/13916#t=suppl

Desk Editors: Soundararajan Raghuraman/Joseph Ang

Typeset by Stallion Press
Email: enquiries@stallionpress.com

Foreword by Liu Yunqi

In 2004, Andre Geim and his student Konstantin Novoselov, two physicists from Manchester University, UK, successfully separated graphene from graphite by transparent tape stripping and characterized its properties. In a matter of six years, the two scientists won the 2010 Nobel Prize in Physics for "their pioneering experiments on the two-dimensional material graphene," a fast achievement in the history of Nobel Prize. They showed the wonders of quantum physics to the world. Not only has their research led to a revolution in electronic materials, but it has also greatly promoted the development of automotive, aircraft and aerospace industries.

From zero-dimensional fullerenes and one-dimensional carbon nanotubes to two-dimensional graphene and three-dimensional graphite and diamond, the discovery of graphene completes the family of carbon materials. As a new two-dimensional carbon nanomaterial, graphene has been in the spotlight since its birth and quickly attracted worldwide attention, sparking the research interest of general scientific research personnel. Known as the "King of new materials," graphene is the thinnest, hardest, most conductive and thermal conductive material known to man. On the one hand, its excellent performance has aroused people's research enthusiasm; on the other hand, it has also set off a wave of application development and industrialization. Graphene has a wide application prospect in composite material, energy storage, conductive ink, smart coatings, wearable devices, new energy vehicles, rubber and health industries. Against the current backdrop of a new round of industrial upgrading and scientific and technological revolution, the advanced material industry will become the cornerstone and precursor of future high-tech industry development, thus exerting a profound impact on the global economy, science and technology, the environment and other fields of development.

China is a country with plenty of graphite resources; it is also one of the most active countries in graphene research and application development. China has become the world's most powerful driving force for the development of graphene industry, taking a dominant position in the global graphene market.

As an advanced material at the strategic frontier in the 21st century, graphene has been developed in China for more than a decade. During this timeframe, it has made gratifying achievements whether in scientific research or industrialization, but at the same time, it also faces a number of bottlenecks and challenges. How to realize the controllable and macro-scale preparation of graphene, and how to develop the function of graphene and expand its application field, are common and key scientific issues in the development of graphene industry in China. In this context, in order to elevate the theoretical basic research of new graphene materials and industrial applications to a new level in China, improve the development system of graphene industry and its large-scale application in many fields, and promote the research system, the development of disciplines, the construction of professional personnel and the training of talents in the field of graphene science and technology in China, a set of high-quality volumes was born. Led by Academician Liu Zhongfan, Professor of Peking University and President of Beijing Graphene Research Institute, *Advanced materials at the Strategic Frontier — Graphene Publishing Project*, a set of 22 volumes, systematically teases out graphene industry chain knowledge from five parts: basic properties and characterization techniques of graphene, preparation technology and measurement standard of graphene, classified applications of graphene, development status report of graphene and popular knowledge of graphene. The series boasts reasonable layout by integrating various relevant points, presenting the ideas clearly presented with sharp focus, with a view to exploring new heights in graphene-based research, tracking the growth of graphene industry, reflecting major innovations and displaying independent intellectual property achievements in the field of graphene. As such, it serves as a good reference for decision-making in major planning of new materials at the strategic frontier in China.

The experts and scholars involved in planning and writing of the series come from more than 20 universities, scientific research institutions and related enterprises in China. Taking a national perspective and positioning themselves at the forefront of scholarship, they have collated and summarized graphene research results in a rigorous academic manner, in order to publish fine works of the time. The series presents readers with sound scientific theories, accurate literature data and a wealth of experimental cases. It provides important guidelines for basic theoretical research and industrial technology upgrading of graphene, leading the majority of scientific and technological personnel to further explore, study and breakthrough more technical problems of graphene. It is believed that the series will become a benchmark in the field of graphene publishing.

I am particularly gratified and grateful that this series of books has been included in the "Thirteenth Five-year Plan" for the National Key Publications and has the full support of National Publishing Fund. I would like to express my gratitude to all my colleagues who participated in the preparation of the series and to East China University of Science and Technology Press. Without your dedication and cooperation in your respective fields of expertise, this set of high-level academic monographs would not have been published successfully.

Finally, as an advisory member of the editorial board, I would like to recommend this series to our readers.

Liu Yunqi
Academician of the Chinese Academy of Sciences
Institute of Chemistry, Chinese Academy of Sciences
Beijing, China
April, 2020

Foreword by Cheng Huiming

Advanced materials at the strategic frontier — Graphene publishing project:

An integrated collection of books on graphene

On October 5, 2010, I attended a cross-strait seminar on new carbon materials in Taiwan, China, and gave an invitation report on "The exploration of Graphene preparation and application." A few hours later, I got a piece of very exciting news for anyone involved in graphene research and development: the 2010 Nobel Prize in Physics was awarded to Professors Andre Geim and Konstantin Novoselov for their pioneering experimental work in the field of graphene.

Carbon is probably the most magical element known to man, something all of us can't live without even for a single moment. The fuel we use is all carbon-based, the food we eat is mostly carbohydrates and the air we exhale is carbon dioxide. Not only that, in nature, pure carbon exists mainly in two forms: graphite and diamond, the former makes Chinese calligraphy possible, while the latter symbolizes beautiful love and happy marriage. Since the beginning of 1980s, carbon has surprised man again and again: At the start of 1980s, scientists used chemical vapor deposition to grow single crystals and thin films of diamond under mild conditions. In 1985, Kroto from Sussex University of UK cooperated with Smalley and Curl from Rice University of America to discover fullerene with a perfect structure and was awarded Nobel Prize in Chemistry in 1996. In 1991, Lijima from NEC of Japan noticed tubular nanostructures composed of carbon and formally proposed the concept of carbon nanotubes, which greatly promoted the development of nanotechnology, thus winning Caffrey Award for nanoscience in 2008. In 2004, Geim and his doctoral student Novoselov isolated graphene materials by stripping graphite with adhesive tape, which quickly sparked research interest of the world. In fact, graphene's structure is no stranger to humans. Graphene is a two-dimensional honeycomb made of a single layer of carbon atoms, the basic building blocks of carbon materials in other dimensions. So, the work on the structure of graphene goes back to the theoretical study of 1940s. In 1947, Walace first calculated the electronic structure of graphene and discovered that it has a strange linear dispersion relation. Since then, graphene, as a theoretical model, has been widely used to describe the structure and properties of carbon materials. However, graphene itself has not been studied and developed as a material.

Graphene material has attracted attention of people in various fields the moment it appeared and rapidly became the star in new material, condensed matter physics and other fields. It has surpassed fullerene and carbon nanotube — the two-star materials already active in the carbon family mainly for three reasons. One is that graphene is relatively simple to make. Geim and his team used a simple and effective mechanical stripping method to separate multi-layer or even single-layer graphene from high quality graphite by tearing with a sticky tape. Scientists then used a similar principle to develop a "top-down" method of stripping for the preparation of graphene and its derivatives, like graphene oxide, or they used a "bottom-up" method of chemical vapor deposition similar to the preparation of carbon nanotubes to grow single and multiple layers of graphene. The other is that graphene has many unique, excellent physical and chemical properties, such as massless Dirac fermion, quantum Hall effect, bipolar field effects, extremely high carrier concentration and mobility, ballistic transport characteristics of submicron scale, and super-large specific surface area, extremely high thermal conductivity, transmittance, elastic modulus and strength. Finally, especially because of these excellent properties, graphene has the potential to gain important applications in many fields such as information, energy, aviation, aerospace, wearable electronics and smart health, including but not limited to new energy battery, high-efficiency heat dissipation film, transparent touch screen, super sensitive sensor, smart glass, low loss fiber, high-frequency transistor, body armor, lightweight high-strength aerospace materials, wearable devices and so on.

Due to its simplest and most perfect two-dimensional crystal structure, massless fermion characteristics, excellent performance and broad application prospects, graphene has given academics and industry much room for imagination, with the potential to lead to breakthroughs in many areas of technology. Major countries in the world have attached great importance to the development of graphene. Many universities, research institutions and companies are committed to the basic research and application development of graphene, hoping to bring significant scientific breakthroughs and market value. Not willing to be outdone, China is the world's most active country in graphene research and application development, ranking first in the world with a very large graphene research and development team. The relevant statistics show that China far outnumbers any other country in the world not only in terms of the number of officially published graphene-related academic papers and the number of graphene-related patents filed and granted in China but also in terms of the number of graphene-related companies in China and the size and variety of graphene products. However, despite 16 years of research and development on graphene, we still face a series of important challenges, particularly high-quality graphene controllable-scale preparation and the expansion of irreplaceable applications.

Over the past 16 years, many countries all over the world have invested huge manpower, material and financial resources in the field of graphene research, development and industrialization. They have made considerable progress in the preparation technology, physical property control, structural construction, application expansion, analysis detection, standard equation and many other aspects, forming a treasure house of knowledge. Although a number of Chinese books on graphene have been published, no attempt has been made yet on a comprehensive, systematic summary, analysis and publication of this treasure house of knowledge, to guide the sustainable development of research and application of graphene in China.

To this end, it is a true blessing that China's major pioneer in the graphene research field and important promoter of the development of graphene in China, Academician Liu Zhongfan, professor of Peking University and founding director of Beijing Graphene Research Institute, personally designed and served as editor in chief of the series "Advanced Materials at the Strategic Frontier — Graphene Publishing Project." The series consists of five parts and a total of 22 volumes, addressing the basic properties and characterization techniques of graphene, preparation technology and measurement standard of graphene, classified applications of graphene, current status report on the development of graphene and popular science knowledge of graphene. It is contributed by Academician Liu Zhongfan, Academician Zhang Jin and other well-known experts in graphene research, application development, testing and standards, platform construction and industrial development. With a comprehensive macroscopic look at graphene, these contributions not only summarize the latest research progress in the field of graphene domestically and internationally, including the contributors' years of painstaking research accumulation and experience, introduce systematically the present situation and development prospect of the industrialization of graphene as a new material, but also present the global graphene industry report and China graphene industry report. In particular, in order to better educate the public about graphene, Academician Liu Zhongfan also took the lead in writing the popular science volume *Q&A: The Magic of Graphene*; this remarkable work blends ingenuity with practicality, thus making the book series a must-read. I show deep respect for them for completing this massive work in their busy schedule, and I believe that their contribution will be an important impetus to the development of graphene field in China and the world at large.

Academician Liu Zhongfan has always stressed that "preparation determines the future of graphene." Here I would like to join him by saying that "the future of graphene lies in applications." I sincerely hope that this book series will help us invent and develop high-quality graphene preparation technology and assist us in expanding the competitive application fields for graphene. With the full cooperation of government, industry, university, research and application, graphene, the newest member of the carbon family that boasts the simplest yet most excellent performance, will become the magic material that sustains human development.

Cheng Huiming
Academician of Chinese Academy of Sciences
Tsinghua-Berkeley Shenzhen Institute
Tsinghua University, Shenzhen
Institute of Metals, Chinese Academy of Sciences
Shenzhen
Shenyang National Laboratory for Materials Science
Shenyang
April 2020

Preface to the Series

Graphene is another legend in the allotrope of carbon family. A superstar across academia and industry today, it is almost a household name. Of course, this is well deserved. As a honeycomb-like two-dimensional atomic crystal consisting of a single layer of carbon atoms, graphene has unrivaled properties. Theoretically, it is the best material for electrical and thermal conductivity, and also the ideal light-weight high-strength material. For this very reason, it attracted worldwide attention once it was launched. Graphene has the potential to create a whole new industry, which will be the locus of competition for future global high-tech industry. This has become worldwide consensus of academic and industrial circles.

Graphene has not been around long. It has been only 16 years since Andre Geim and his student Konstantin Novoselov published the first hot spot article on graphene in the American journal *Science*. It shall be noted that previous studies of graphene have accumulated a lot in a span of 60 years. So, we can't draw the simple conclusion that graphene was discovered in 2004 by Andre Geim and his student Konstantin Novoselov. That said, the pioneering contribution of the two scientists to "graphene craze" is unquestionable. It is they who have succeeded for the first time in studying the unique properties of true "graphene material," by using the simple transparent tape stripping method. This experimental approach to isolating graphene has given more scientists the opportunity to carry out related research, leading to a continuous graphene research boom. On October 5, 2010, these two trailblazers were awarded Nobel Prize in Physics, only six years after their first published paper on graphene, "the Carbon Element that forms the basis of all known life on Earth, has once again shocked the world," Royal Swedish Academy of Sciences said in the Nobel press release that year.

From experimental samples in the hands of scientists to graphene commodities in the lives of ordinary people, the advancement of graphene advanced material industry is undoubtedly the fastest ever in history. Europe is the cradle of new graphene materials, and Europeans also want to become the leader in graphene advanced material industry. An important move is the launch of "EU's Graphene Flagship programme," which will invest 100 million euros a year for 10 years in a row from 2013, to accelerate industrialization of new graphene materials through joint efforts of scientists, engineers and entrepreneurs. University of Manchester in UK is the birthplace where new graphene materials emerged; it is also the first place in the world to set up a dedicated graphene research facility. In March 2015, UK's National Graphene Research Institute (NGI) set sail at University of Manchester. In December 2018, University of Manchester also established Graphene Engineering Innovation Center (GEIC). The succession of actions which aimed at both basic and applied research showcased commitment of the university to leading graphene industry. Of course, competition in graphene advanced material industry is fierce. While the United States and Japan strive to emulate the United Kingdom, South Korea and Singapore are well on their way. According to incomplete statistics, there are already 179 countries or areas in the world which have joined graphene research and industry competition.

China's graphene research started very early but has managed to keep pace with the rest of the world. Most colleges and universities all over the country with science and engineering departments are more or less working on graphene research. The same is true of all the research institutes under the jurisdiction of Chinese Academy of Sciences, the national team of scientific and technological innovation.

With the world's largest graphene research team and its robust innovative energy, the total number of graphene-related academic papers contributed by Chinese scholars has been the highest in the world since 2011 and by a wide margin. By March 2020, the contributions from China Mainland had reached 101,913, accounting for 32% of the world's total. It is important to note that China not only leads in terms of statistics but also lots of innovative and leading results. Super clean graphene, super graphene glass and carbon fiber are typical examples, to name but a few.

China keeps pace with or even outpaces rest of the world in graphene industry. Statistics show that as early as 2010, the number of formally registered enterprises to carry out graphene-related business reached 1,778. By February 2020, that number jumped to 12,090. Naturally, for graphene high-tech industry, the competition for intellectual property rights is very fierce. Since the beginning of 21st century, the issue of intellectual property rights has been paid more attention than ever before; this is evident in the field of graphene as a new material. By the end of 2018, the total number of graphene-related patent applications worldwide amounted to 69,315, among which those from China Mainland reached 497, coming out first at a percentage of 64. Therefore, statistically speaking, China's graphene research and industrialization process is undoubtedly leading the world. Admittedly, statistics can only tell you part of the truth, but they can also mask some important truths. This, of course, is not confined to the field of graphene as a new material.

China's "graphene craze" has been going on for nearly a decade, even to the point of mania, a phenomenon rarely seen in other countries and areas of the world. Especially at the height of "graphene gold rush" a few years back, places all over the country raced to build "graphene industrial parks," "graphene towns" and "innovation centers in graphene industry" and even set up graphene research institutes in towns and counties, such flurry of graphene-related activities is a bit like "Great steelmaking movement" in 1950s. To be fair, China's graphene industry is the fastest growing in the world. It has one of the largest industrial forces in the world and even attracted overseas "gold diggers," including the two graphene Nobel laureates, among others. There is also no denying that some unhealthy factors exist in the development of China's graphene industry. With everyone eager to jump on the bandwagon, a lot of simple duplication of construction and low-level competition have ensued. Take the production of graphene as an example; in 2018, annual output capacity of the powder material reached 5,100 tons and that of CVD thin film reached 6.50 million square meters, more than the sum of other countries and areas, thus causing the issue of overcapacity. On January 30, 2017, in an interview with ThePaper.cn, I explicitly expressed my concerns about the current development of China's graphene industry. Soon afterward, it received high attention and instructions from General Secretary Xi Jinping. The relevant departments have followed General Secretary Xi's instructions and made a nationwide survey of the development status of graphene industry. Now, three years later, it can be said that the situation has changed. With the public's better understanding of the new materials of graphene and industrialization practice from laboratory to market, China's "graphene fever" has cooled down and people are getting more rational.

It was in this context that the voluminous graphene book series was born. Sixteen years have elapsed from 2004 until now. Whether it is the basic research of graphene, or the industrialization practice of graphene materials, people have more first-hand information and are more likely to have a comprehensive, scientific, rational understanding of graphene materials. To review past successes and failures is to move toward the future in the right direction. For the emerging graphene industry, the significance of the series is self-evident. In fact, dozens of graphene-related books have been published worldwide, with no lack of classic works. This series has a different positioning: it hopes to comprehensively summarize the knowledge accumulation related to graphene, reflect the latest research progress in the field of graphene both domestically and internationally and present industrialization status and development prospect of new graphene materials. In particular, it hopes to fully demonstrate the contribution of Chinese people to the field of graphene. Like a marathon, the series took nearly five years from planning to completion. Without creativity, persistence and great patience of East China University of Science and Technology Press Project Team, publication of this series would have been unthinkable. Their persistence in reaching their goals moved me so much that I undertook this glorious yet arduous task. This persistent spirit also ran through the whole process of compilation. Every author is committed to producing their best possible work.

The series consists of 22 volumes and is written by more than 20 authors. They are all leading experts in the field of graphene or experts in standard graphene metrology and industry analysis.

Therefore, we can guarantee the professionalism and authority of the series from the source. The series is divided into five parts, covering basic properties and characterization techniques of graphene, preparation methods of graphene materials and their applications in different fields, as well as a summary of standard of measurement and test for graphene products. At the same time, two latest industry studies reports elaborate on current situation and future trends of development of the world's graphene industry. In addition, the series also provides *Q&A: The Magic of Graphene*, a popular science reading material for graphene fans, to clarify and demystify graphene-related questions collected extensively from the public.

The specific contents of each volume and the division of writing are as follows:

Vol. 01. *Structure and Basic Properties of Graphene* (Liu Kaihui); Vol. 02. *Characterization Technology of Graphene* (Zhang Jin); Vol. 03. *Study on Raman Spectra of Graphene-based materials* (Tan Pingheng); Vol. 04. *Preparation Technology of Graphene (Peng Hailin)*; Vol. 05. *Chemical Vapor Deposition Growth of Graphene* (Liu Zhongfan); Vol. 06. *Preparation Method of Powder Graphene Material* (Li Yongfeng); Vol. 07. *Quality Infrastructure of Graphene and Related Two Dimensional Materials: Metrology* (Ren Lingling); Vol. 08. *Electrochemical Energy Storage Technology of Graphene* (Yang Quanhong); Vol. 09. *Graphene Supercapacitors* (Ruan Dianbo); Vol. 10. *Graphene Microelectronics and Optoelectronic Devices* (Chen Hongda); Vol. 11. *Graphene Films and Flexible Optoelectronic Devices* (Shi Haofei); Vol. 12. *Graphene Membrane Materials and Environmental Applications* (Zhu Hongwei); Vol. 13. *Graphene-based Sensor* (Sun Litao); Vol. 14. *Graphene Macroscopic Materials and Applications* (Gao Chao); Vol. 15. *Graphene Composites* (Yang Cheng); Vol. 16. *Graphene Biotechnology* (Duan Xiaojie); Vol. 17. *Chemistry and Assembly Technology of Graphene* (Qu Liangti); Vol. 18. *Functionalized Graphene Materials and their Applications* (Zhi Linjie); Vol. 19. *Graphene Powder Materials: From Basic Research to Industrial Applications* (Hou Shifeng); Vol. 20. *Global Graphene Industry Research Report* (Li Yichun); Vol. 21. *Research Report on China's Graphene Industry* (Zhou Jing); Vol. 22. *Q&A: The Magic of Graphene* (Liu Zhongfan).

This series covers all aspects of the advanced material: graphene; each volume is a separate book in its own right, boasting a strong sense of systematicness, knowledge, professionalism and immediacy. It is a crystallization of the authors' research, wisdom and effort and can meet different needs of general readers as a reference book.

I hope that the publication of the series can facilitate China's graphene research and the healthy development of China's graphene industry. On the occasion of the publication of this book series, I would like to express my sincere thanks to all the authors for their hard work. At the same time, I would like to express my highest respect and sincere gratitude to East China University of Science and Technology Press for their full commitment from start to finish. All the deficiencies and omissions remain ours.

Liu Zhongfan
Mo Yuan
March 2020

Preface to the Volume

Dmitri Ivanovich Mendeleev once said, "no measurement, no science." In many cases, breakthrough of the bottleneck of science and technology depends on the solution of key measurement problems. The credibility of measurement results, or the quality of the measurement, is a matter of great concern. With advancement of globalization and economy developments, measurement accuracy directly affects economic interests of the state and enterprises and thus is in dire need of metrological support. Metrology is the science of measurement. It covers all aspects of measurement theory and practice, regardless of uncertainty of measurement, or which field of science and technology measurement is carried out. Metrology is a technical subject in classification, but it is closely related to social economic life and serves as the fundamental institution of any political system. For this reason, the role of metrology in advancing basic science research and industrialization needs to be further elaborated, so as to be understood by more people.

Graphene is a two-dimensional crystal made of carbon atoms with just one atom thick. As a new material with excellent performance, graphene has attracted the attention of scientific community, business community and governments of various countries. Especially in graphene industry, China has the world's largest production capacity, but graphene material preparation as well as back-end application R&D and industrialization process need to be further strengthened. At present, graphene industry is in the synchronous phase of basic research and industrialization, and this synchronous phase is the optimal timing for metrology to play its role.

Funded by National Key R&D Program of China: "Technology research, integration and application of metrology, standardization and conformity assessment of Graphene and other carbon-based nanomaterials" (No. 2022YFF0608600 & 2016YFF0204300), our team in National Institute of Metrology, China conducts research, integration and application demonstration from three national quality infrastructures (NQI), namely graphene materials metrology, standards and conformity assessment. Encouraged by the progress of National Key R&D Program and the development of graphene material industry in China, we write this book *Quality Infrastructure of Graphene and Related Two-Dimensional Materials: Metrology*.

Research fellow Ren Lingling is responsible for overall design, content deployment and review of the book, while Gao Huifang and Yao Yaxuan are responsible for final editing and reviewing of the manuscripts. This book consists of seven chapters. The first two chapters introduce basic concepts of metrology and material metrology, as well as graphene material measurement, standard layout design and framework. The last five chapters mainly introduce metrological techniques on structural quality evaluation of graphene materials, such as Raman spectroscopy, X-ray diffraction, atomic force microscopy and electron microscopy, to authentically judge true or fake of this material. But for a certain kind of material, the mentioned technologies for the whole product property authenticity judgment are only a drop in the ocean, so to speak. Just as Fig. 28 in Chapter 2 shows, we try to lay out metrological technology framework for quality evaluation of graphene materials, such as X-ray photoelectron spectroscopy, inductively coupled plasma mass spectrometry and Fourier transform infrared spectrometer. By standardizing measurement methods and tracing measurement results of these metrological techniques, comparable measurement results can be obtained. And only in this way can we acquire internationally accepted technical language, to have the world endorse the measurements from each lab. This is the very significance of graphene metrology.

Industrial application of graphene materials needs the support of NQI system. We must establish measurement standard equipment, standard materials, and standard methods or technical specifications for graphene materials. At present, the development of China's graphene industry is at a critical juncture. By solving the accuracy and consistency of measurement through metrological technology research of measurement methods, establishing and following universal standardization system, the final product can be ensured to achieve designed quality and product quality control. In future, China will continue to promote comparative tests between different laboratories and different methods worldwide, facilitate development of standards, and promote standardization, mass production and sustainable and healthy developments of graphene industry.

Basic research and application of graphene enjoys a very fast development. With mushrooming of knowledge and findings, the amount of literature increases exponentially. Please don't hesitate to correct me should you find any mistakes or flaws in this book.

This book has seven chapters. The first two chapters introduce basic concepts of metrology and materials metrology, as well as planning and frameworks of graphene materials' metrology and standard. The following five chapters mainly introduce metrological technologies about true or false judgment in structural quality evaluation of graphene materials. Chapter 3 introduces metrological technology of graphene materials by Raman spectroscopy. Chapter 4 introduces metrological technology of graphene material crystal structure by X-ray diffraction. Chapter 5 introduces measurement of graphene thickness by atomic force microscopy. Chapter 6 introduces metrological technology of graphene materials by electron microscope. Chapter 7 introduces measurement technology of chemical composition of graphene powder. The Appendix supplements traceability study of X-ray diffractometer and calibration of grazing incidence X-ray reflection instrument on film thickness measurement.

Ren Lingling
March 2020

About the Editors

REN Lingling, Ph.D., is a Research Fellow and National First-Level Registered Metrological Surveyor who graduated with a doctorate in 2001 from Beijing University of Chemical Technology. As the head of Advanced Materials Measurement Lab, she is in charge of lab's administration and research on materials measurement. Her work focuses on standardization of advanced material measurement such as graphene-related 2D materials. She was the principal project leader of two projects under the National Quality Infrastructure (NQI) Program of National Key Program of Science and Technology. She is the secretary-general of National Technical Committee of Material and Nano-technology Metrology, the delegate of National Technical Committee for Nanotechnology, the co-chair of VAMAS/TWA41, the former chair of APMP/TCMM, the delegate of ISO/TC229 and IEC/TC113, the delegate of National Technical Committee for Nanotechnology Standardization (SAC/TC279), and the secretary-general of Foundation and Generic Standardization Field Committee of Chinese Standards for Testing and Materials (CSTM/FC00). Over twenty certified reference materials have been released, and six interlaboratory comparisons under APMP/TCMM and VAMAS are underway. Three honors were obtained, and over fifty articles with a total of over 100 citations were published in international journals (SCI).

About the Contributors

Bu Tianjia, Ph.D., is an Associate Researcher who graduated from State Key Laboratory of Supramolecular Structure and Materials, Jilin University, for a doctorate degree, majoring in Polymer Chemistry and Physics. She received government-sponsored study in University of British Columbia for three years during the Ph.D. research. Her research work focuses on effective measuring methods of graphene materials and metrological study of fluorescence measurement. She was responsible for and participated in six scientific research projects, including project of National Key Research and Development Program of China, National Natural Science Foundation of China, and State Administration for Market Regulation Science and Technology Plan. She also led and participated in two international comparisons, and developed ISO standard for GO flake thickness using AFM, as well as national standard and group standard. She has published more than ten papers in international journals.

Dong Guocai, Ph.D., is a Senior Engineer who graduated from Peking University with a bachelor's degree and Leiden University in the Netherlands with a doctorate degree majoring in Physics. Since 2012, He has been in Jiangnan Graphene Research Institute. He is mainly engaged in the development of graphene manufacturing equipment, as well as standardization research in the field of graphene. He was responsible for and participated in the establishment of two national standards, one industry standard, five local standards in Jiangsu Province, one group standard, one national certification standard material, and the construction of Jiangsu Province graphene material standard system. He has written two books and published thirteen papers.

Gao Huifang is a Senior Engineer, the national first-level registered metrological surveyor, and national first-level metrology standard evaluator. She graduated from Sichuan University in 2010 and then came to National Institute of Metrology in China. She is mainly working on metrology study of nanostructure for advanced materials, such as semiconductor materials and carbon-based nanomaterials including graphene. She is in charge of two subprojects under national program and two institute-level projects, and has participated in two national projects and one institute project. She was responsible for and participated in the establishment of three national measurement standards, nineteen national certified reference materials, one national standard, one group standard, and three national metrology technical specifications. She was in charge of one domestic comparison project and participated in four international comparisons, and published over ten papers. She was awarded one first prize (4th participant) and one third prize (6th participant)

from China Association for Instrumental Analysis, and one third prize (2nd participant) from Chinese Society for Measurement.

Ge Guanglu is a Professor at National Center for Nanoscience and Technology (NCNST) and obtained his Ph.D. from Columbia University in 2001 and subsequent postdoctoral training at UCLA and Caltech. He joined NCNST in 2005 and since then has focused the research efforts on standardized testing methods and reference materials for nanomaterial characterization and nanoscale interface properties. He has published over 80 papers in journals including Advanced Materials and Small. He received the IEC 1906 award in 2018.

Prof. Ge is actively involved in nanotechnology standardization. He is now the convenor of ISO/TC229/WG4 (Nanomaterial specification) and the deputy director of Technical Committee on Nanotechnology of Standardization Administration of China (SAC/TC279). Prof. Ge has led or co-led the drafting of seven ISO/IEC international standards and sixteen national standards. Specifically, by collaborating with domestic industry, his team developed a series of physicochemical testing standards for electrode nanomaterials and optical characterization of quantum dots.

Li Jinlong graduated from Anhui University of Technology, majoring in Materials Science and Engineering. He has published four papers in international journals (SCI) during 2010–2013. He is mainly working on measurement study of graphene at the Growth Mechanism Research Laboratory, Jiangnan Graphene Research Institute (2013–2016). He is currently positioned at the Changzhou branch center for evaluation and inspection of JSMPA.

Li Xu, Ph.D., is an Associate Research Fellow and graduated from Central South University in 2014, majoring in Materials Science and Engineering. He is now working in materials metrology laboratory/Center of Advanced Measurement Science, mainly engaged in the study of nanoscale measurement and advanced materials, and electron microscope metrology. In 2016, he took part in National Key Research and Development Project "The study, integration and application of NQI technologies of graphene and other carbon-based nanomaterials." In 2017, he took part in the National Key Research and Development Project "The metrological standard study of key characteristic parameters of advanced functional materials." As the project leader, he was in charge of one National Quality Infrastructure projects and completed one National Science Foundation Youth Fund project. He has published 30 papers in international journals.

Liu Renxiao, Ph.D., is a Senior Engineer at National Center for Nanoscience and Technology (NCNST). She obtained her Ph.D. degree from China Petroleum University in 2008 and subsequently joined NCNST. Her research areas focus on the industrial application standardization of nanotechnology and typical nanomaterials, especially fluorescence nanomaterials, graphene, carbon nanotubes, colloidal particles in optoelectronic, and electrothermal application area. She has led to complete over 20 ISO and IEC nanotechnology international standards and Chinese national standards.

Tian Guolan is an Engineer at CAS Key Laboratory of Standardization and Measurement for Nanotechnology. She graduated from Beijing University of Technology in 2015. Her main research area is the controlled synthesis and measurement standards development of nanomaterials, including the study of chemical composition measurement of graphene materials. She participated in two key research projects and published scientific papers and technical standards.

Xu Peng, Ph.D., is a Senior Engineer, graduated from Beijing University of Technology in 2018, and majored in Applied Chemistry. She has been engaged in surface analysis work since 2010 and is committed to the material surface structural analysis and reaction mechanism research.

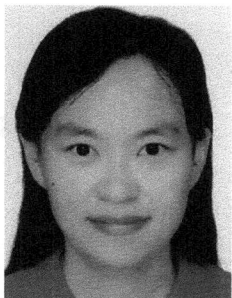

Yao Yaxuan, Ph.D., is an Associate Researcher. She is a member of CCQM/SAWG (an international metrology organization). She has graduated from University of Science and Technology of China with a bachelor's degree and University of Illinois at Urbana-Champaign in US with a doctorate degree. Since 2013, she has been working in National Institute of Metrology in China. She is mainly working on metrology study of spectroscopy instrument including Raman spectroscopy, and accurate measurement study of carbon-based nanomaterials, including graphene. As the project leader, she is in charge of two National Quality Infrastructure projects and one National Science Foundation Youth Fund project. She was responsible for and participated in the establishment of one national measurement standard, eight national certified reference materials, two national standards and their English versions, and two association standards. She has published more than 10 papers with citations over 200 times.

Zhang Xiaomin is a Standardization Senior Engineer. She graduated from Changzhou University, majoring in Organic Chemistry. She is mainly engaged in standardization work in the field of graphene. As a key researcher, she was responsible for and participated in the establishment of one national standard, three industrial standards, five local standards of Jiangsu Province, one group standard, and one national-certified standard material in the field of graphene, and the construction of Jiangsu graphene material standard system. She has published five papers.

Contents

https://doi.org/10.1142/9789811295713_0001

Chapter 1

Metrology and Materials Metrology

Lingling Ren

Graphene opens up a new, delicate world of materials. Since its discovery, research work on graphene and other two-dimensional materials is in full swing. Given its good light transmittance, high thermal conductivity, high electron mobility, low resistivity, high mechanical strength and other excellent performances, graphene is expected to lead to the rapid development in a new generation of information technology, new energy, high-end equipment manufacturing and other fields if a major breakthrough can be made in large-scale preparation and its application. More and more attention has been paid to the research and industrialization of graphene in the world. Meanwhile, developed countries generally adopt the materials development strategy of "Developing, reserving, and applying a batch of materials." China has continuously strengthened its policy support in this regard. It has included graphene in its major project *Outline of the Thirteenth Five-Year Plan for National Economy and Social Development of the People's Republic of China*, incorporated graphene in the major field *Made in China 2025*, and issued *Several Opinions on Accelerating the Innovative Development of the Graphene Industry* in 2015. Experts predict that graphene is expected to enter the one trillion-tier market in 2020.

In order to better support basic research, industrial transformation of research achievements and high-quality development of graphene industry in China, the national key R&D program of China has provided fundings for the key projects "Research, integration and application of metrology, standardization and conformity assessment of graphene and related carbon-based nanomaterials" (Project number: No. 2022YFF0608600 & 2016YFF0204300, hereinafter referred to as Graphene NQI Projects). Before elaborating on the progress of metrological technologies of graphene materials, this book introduces metrology, materials metrology, international comparison and other concepts. It first gives a general overview of the subject matter and then presents a detailed explanation of the specific points so that readers can have a clearer understanding of metrological technologies of graphene materials.

1.1 Overview of Metrology

Metrology is an activity aiming to achieve the unity of units and accurate and reliable quantity value. People's lives are inseparable from metrology. When we weigh the goods, start the taximeter, estimate the house area, settle the bill, etc., we are literally using metrology in our daily routines. In China, ancient metrology started more than 4,000 years ago. "At the mention of metrology, many people will think of the weights and measures unified by Qin Shi-Huang (First Emperor of Qin Dynasty, 259–210 BC). In fact, in ancient China, Yellow Emperor 'set five measures of volume, quantity, weight, length and number,' and 'Yu the Great's voice was the law, his behavior the standard, he demonstrated these qualities in such a proper manner that they became the yardstick for his people,' proving that China has a history of over 4,000 years in its legal measurement system, 2,000 years earlier than the weights and measures unified by Qin Shi-Huang," said Ai Xuepu, one of the main organizers of Tianjin Institute of Metrology, Supervision and Testing. A lot of people used to regard measurement as "merely ruler, bucket and scale," which is actually a misunderstanding lingering for over 2,000 years. In a real sense, the emergence of primitive metrology went through a long process in the history of China, starting from simple counting measurement like "keeping records by tying knots" and "keeping time by marking on the wedge" to using human parts as the instrument of measurement like "taking the width of a palm as *chi*, the volume of a handful as *sheng*, the volume of two handfuls as *yi*, and the length of a footstep as *mu*" and "using Yu the Great's standard as the yardstick for his people." As is recorded in *Zi Zhi Tong Jian (History as a Mirror)*, "Yellow Emperor ordered his historiographer Lishou to set rules for counting and calculating, to accumulate the surplus and check the total, hence the unified standard for measuring volume, quantity, weight, length and number," proving that our ancient metrology formed and produced the laws, weights and measures which reflected the margin of error during Yellow Emperor's period of the paternal clan society, over 4,000 years from today. The well-documented metrology dates back to prehistoric times when Qin Shi-Huang unified the weights to measure length, a decimal scale of *hu*, *dou* and *sheng* to calculate volume, and a decimal scale of *shi*, *jun*, *jin*, *liang* and *zhu* to measure weight. Qin Shi-Huang conquered the six kingdoms and unified the country, issued decrees to unify the system of weights and measures throughout the country in legal form, and engraved or cast the command on volume and weighing instruments as proof of use. The court manufactured and issued standard instruments of measurement as the yardstick for local productions and certifications, and certified these instruments once a year. Qin Dynasty unified weights and measures and laid the foundation for the basic system of weights and measures in China's feudal society which continued in use for over 2,000 years. The unification of weights and measures has greatly facilitated nationwide exchange of commodities and economic activities, and provided a strong guarantee for normal operation of state machinery and social activities. Even the modern legal system of metrology includes unit system and instrument manufacture system similar to that of Qin Dynasty. Mao Zedong (1893–1976, the first Chairman of People's Republic of China) once said in the seven-character octave, "On *Reading feudalism*: to Mr. Guo Moruo," that "all the past hundreds of dynasties practiced the politics and law of Qin Dynasty." Thus, we can see that the legal system of weights and measures of Qin Dynasty had a far-reaching impact on future generations.

The need to establish common standards of measurements around the world was made clear for the first time at the First World's Fair (hosted by the Royal Society of Arts) in London, UK, in 1851. This is the first international fair to attract exhibits from all over the world. At the exhibition, technical specifications for products, machinery, equipment and other exhibits from different original countries were rather confusing, with units in English units, metric units or other units. That complicated the task of judging and selecting the winners of many awards, which formally triggered the promotion of adopting an international uniform measurement system. In 1853, the Memorandum of Council of British Academy of Arts recorded the following: "Practice shows that the unification of money and metrology is crucial to business activity. As far as metrology is concerned, this will also greatly enhance scientific researches at the same time. Hence, the governments must carry out necessary research on metrology to achieve the conversion of money and measurement to decimal system in the best way possible. Neighboring countries shall take steps to promote a unified system around the world through consultations." Based on the growth of international manufacturing trade and the rapid development of science and technology in the mid-19th century, on May 20, 1875, representatives from 17 member countries signed *the Metre Convention* in Paris, France, and formally agreed to promote a unified international unit of measurement and physical measurement, marking the establishment of Bureau International des Poids et Mesures (BIPM). In honor of this great occasion, 21st General Conference on Weights and Measures (1999) announced that May 20 is World Metrology Day.

Thus, since the Age of Enlightenment, a dynasty, a country and even the international measurement field have been committed to establishing a "universal" measurement system. International System of Units (SI) is such a globally recognized measurement system. The seven basic units of SI are Second, Meter, Kilogram, Ampere, Kelvin, Moore and Candela. These basic units can represent measurement results in any field, and the results are comparable and consistent across the globe.

These unit benchmark definitions are initially reproduced physically, for example, "Meter" was originally defined by a metal rod just one meter long, calling it International Prototype Metre (IPM) (Fig. 1.1). Prior to November 16, 2018, "Kilogram" was the last SI basic unit still defined by a physical object, calling it International Prototype Kilogram (IPK) (Fig. 1.2), the only kilogram measurement reference. IPK is placed in an eight-meter-deep strictly controlled underground vault at BIPM headquarters in Sevres, France, untouched by human hand for the past 143 years. Like a unique Matryoshka doll, this column is housed in a triple-glazed bell jar.

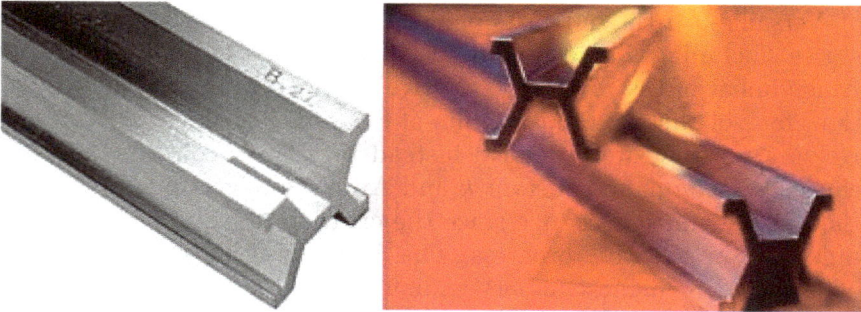

Fig. 1.1.   International prototype of metre.
Source: http://www.bipm.org and Wikipedia.

Fig. 1.2.   International prototype of kilogram.
Source: http://www.bipm.org.

To enter the room, three different keyholders must open the door at the same time.

But these objects can change over time or with the environment and cannot meet the needs of measurement accuracy in today's scientific research and technical application. On November 16, 2018, the 26th annual General Conference on Weights and Measures was held at Versailles on the outskirts of Paris, France, at which 53 member states, including China, unanimously passed Resolution 1 on "the Amendment of the SI" by a collective vote. According to the resolution, four of the seven basic units of SI, namely Kilogram, Ampere, Kelvin and Mole were defined instead by Planck constant h, basic charge e, Boltzmann constant k and Avogadro constant N_A. This is the most significant change since the creation of SI in 1960 and is a milestone in scientific progress. This new definition reconstructs SI on our current highest understanding of natural laws, that is, using the laws of nature to establish measurement rules, associate measurements at atomic and quantum levels with those at macroscopic level, to fulfill the common aspiration of Metre Convention — to provide a universally applicable benchmark for global measurement. At the same time, it eliminated the association between newly defined SI units and those basic units based on the definition of a physical reference and confirmed what Maxwell had predicted at British Association for Advancement of Science in Liverpool in 1870: standards of measurement must sooner or later be defined in terms of immutable constants of nature.[1]

The new SI will meet the following conditions (Fig. 1.3):

(1) transition frequency of hyperfine energy levels in the ground state of 133 Cesium Atom: $\Delta v_{cs} = 9{,}192{,}631{,}770\,\text{Hz}$,
(2) the speed of light in a vacuum: $c = 299{,}792{,}458\,\text{m s}^{-1}$,

Fig. 1.3.   International unit.

(3) Planck constant: $h = 6.62607015 \times 10^{-34}$ J · S,
(4) basic charge: $e = 1.602176634 \times 10^{-19}$ C,
(5) Boltzmann constant: $k = 1.380649 \times 10^{-23}$ J · K^{-1},
(6) Avogadro constant: $N_A = 6.02214076 \times 10^{23}$ mol^{-1},
(7) luminous efficiency of monochromatic radiation at 50×1 z frequency: Kcd $= 683$ lm · W^{-1}.

1.2 Metrology and Measurement

1.2.1 Measurement

The essence of science is measurement. People want to communicate with each other by comparing quantitative results of real-life behaviors without regard to the specific field. Quantifying the results of these behaviors requires some important data, like how big, how much, how heavy, how fast, how strong, what color, what density, etc. These data and information must be transmitted by "measurement." Measurement is the use of data to describe observed phenomena in accordance with certain laws, that is, to quantify things. Measurement is the process of quantifying non-quantifiable objects, enabling the transformation from the qualitative to the quantitative in people's understanding of the attributes of objects, materials and natural phenomena, in order to enhance credibility and scientificness of the law of nature. Section 4.1 of "Terms and definitions of generic metrology," *The National Metrological Technical Code of People's Republic of China* (JF1001-2011), defines "Measurement" as "the process of obtaining and reasonably assigning one or more values to a measurement by experiments." This definition includes three layers of connotations: (1) Measurement is an operation. This operation could be simple and does not require high-accuracy actions, like taking your temperature at home, taking your blood pressure or filling a glass (liter) of beer. Or it could be a complex scientific experiment, like quantum transmission measurement, land area measurement, nanomaterial structure measurement and so on. (2) The whole process emphasized here starts with the clarification or definition of measurement, including the selection of measurement principles and methods, the selection of measurement standards and equipment, and the controlling of critical values through experimentation and calculation, until obtaining a measurement result with appropriate uncertainty. (3) The purpose of measurements is to determine the quantitative value, which is usually expressed by multiplying a number by a measurement unit. There is no specified measurement scope and corresponding uncertainty, so the measurement is applicable to many aspects and fields.[2]

With globalization and the development of China's national economy, whether the measurement results or the measurement quality is reliable is a matter of great concern. The accuracy of measurement may have a direct impact on economic interests of the state and enterprises. As a simple example, the volume flow rate is used as the settlement basis in the import of natural gas. If the volume flow of natural gas in the pipeline is not measured correctly, it will surely cause one party to pay more expenses or the other party to supply more gas, and may even cause economic disputes. The quality of a measurement is often an important factor in the success or failure of scientific experiments. The discovery of graphene, for example, could only exist as a theoretical prediction because it was impossible to measure such a thin two-dimensional material until it was successfully separated from graphite by micromechanical stripping in 2004 by physicists Andrea Geim and Constantin Novosyolov of the University of Manchester, UK, and confirmed by atomic force microscopy (AFM) and electrical measurements that it can exist alone.[3] Research on graphene has thus picked up, and the pair was also jointly awarded Nobel Prize in Physics in 2010. Measurement results are the basis for evaluation of scientific research achievements, as well as criterion for product inspection, conformity assessments, judicial ruling and so on. How can we say that an achievement has reached the international advanced level? How can we say that a product is unqualified?

The measurement results are their rules. The conclusion drawn from measurement results is also an important basis for government management decision-making, such as making decisions on the direction of graphene industry, space satellite launch window and smog control measures. Wrong or unreliable measurements will lead to wrong decisions or shake the confidence of decision-makers, incurring loss because of delay.

1.2.2   Metrology

Dmitri Ivanovich Mendeleev once said, "No measurement, no science." In many cases, breaking through the bottleneck of technical progress depends on the solution of key measurement problems. The essence of measurement is the reference standard to be measured.[4] Science is not a local activity, for the laws of nature are universal, the findings in London are equally valid in Paris, Washington or on the moon. To hold an international dialog on scientific issues, there must be a universal scientific language and the basic principle of describing scientific theories and experimental results. The most common scientific language is mathematics, such as Newton's law and Einstein's field equations $E = mc^2$ that we have learned. Math is undoubtedly the same all over the world. The common language for describing scientific theories and experimental results is "measurement," and that common language requires common reference standards. That was the purpose and meaning of BIPM when it was first established.

Section 4.2 of "General metrological terms and definitions" in *The National Metrological Technical Code of People's Republic of China* (JF1001-2011) defines "Metrology" as "the activities to achieve unity of units, ensure accurate and reliable measurand values." The activities include scientific and technical, legal and administrative activities. Metrology was known as weights and measures in history; the main instruments were the ruler, bucket and steelyard; the weight of a steelyard was called power, which is still used to represent the rule and fairness of the law. In English, the word "ruler" means both the instrument of measuring length and the person who holds a dominant position. All these show that metrology is an activity symbolizing rights and justice.

The purpose of a measurement is to determine the measurand value and ultimately meet the needs of society. Therefore, the values determined for measurand at different times, places by different operators, with different instruments shall be comparable. Only when the selected unit follows the uniform criteria, and makes the obtained measurand value have necessary accuracy and reliability, can this comparability be ensured. Metrology is one of the important means to achieve this goal; in this sense, metrology can be broadly regarded as the process of qualitative analysis and quantitative confirmation of "quantity."

Metrology is also a discipline; it is the science of measurement. Metrology covers all aspects of theory and practice of measurements, regardless of the uncertainty, or in which field of science and technology the measurement is made. In terms of its development as a discipline, metrology is a part of physics which has later formed a comprehensive science to study the theory and practice of measurements with the expansion of application fields and contents. Especially in science, metrology is closely integrated with national laws, regulations and administration, an integration rarely seen in other disciplines. In a narrow sense, metrology is the science of measurement and its applications, which is a standardized measurement associated with the confidence and uncertainty level of the result.

Metrology has been classified in different ways. There is an international tendency to divide metrology into scientific metrology, engineering metrology and legal metrology, representing respectively the metrological basis, metrological applications and social undertakings in which the government plays a leading role. At this point, the discipline of metrology is simply called metrology.

Scientific metrology refers to the scientific research of metrology which is basic, exploratory and pioneering. It often uses the latest scientific and technological achievements to precisely define and implement metrology units, and provide a reliable basis of measurements for the latest scientific and technological developments.

Scientific metrology is the main task of national metrology research institutions, including research on the metrology units and the unit system, the measurement standard instruments, physical constants and precision measurement technology, characteristic parameter and accurate measurement methods, quantity traceability and quantity transfer system, and laboratories comparisons and measurement uncertainty.

Engineering metrology is also called industrial metrology, which refers to the practical measurements in various projects, industries and enterprises, such as energy or material consumption, process monitoring, and product quality and performance measurement. Engineering metrology involves a wide range of areas. With the increase of technical content and complexity of products, engineering metrology has become an indispensable aspect in production process control, to ensure consistency and interchangeability in economic and trade globalization. In fact, the ability of engineering measurements is an important component of a country's industrial competitiveness, particularly in the high-tech-based economic structure.

Legal metrology is the metrology related to the work of legal metrology departments, involving statutory requirements for basic units, standard instruments, measurement methods and measurement laboratories. Legal metrology receives mandatory management from the government or authorized agencies in accordance with legal, technical and administrative requirements, the purpose of which is to stipulate and secure, by statute or contract, the integrity and reliability of measurement work related to trade settlement, security, healthcare, environmental monitoring, resource control, social management, etc. All of these involve public interest and national sustainable development strategy.

1.2.3 Metrological Characteristics

With the development of science and technology, economy and society, the content of metrology has expanded from the original seven fundamental physical units, namely length, mass, time, current, temperature, matter quantity and luminous intensity, to high-tech fields, such as biology, medicine, environmental protection, new materials, information, software and so on. The metrological characteristics depend on measurement fields, i.e. scientific and technological, legal and management activities aiming to achieve unity of units as well as accurate and reliable quantity value. Metrology is marked by accuracy, consistency, traceability and legality, but measurement does not have to boast the above-mentioned four metrological characteristics.

Accuracy refers to the consistency of the measured result with the true value measured. Since there is actually no such thing as a completely accurate measurement, when a quantity value is given, the range of uncertainty must be given to suit application purposes or actual requirements. Otherwise, the quality of the measurement cannot be judged nor can the quantity value have its full practical value. The so-called accuracy of quantity value means the accuracy within a certain degree of uncertainty, error limits or allowable error range.

Consistency means that based on a uniform unit of measurement, the measurement results shall be consistent within a given range, no matter when, where, by what means, with what measuring instruments and by whom, as long as the relevant requirements are met. That is to say, the measurement shall be repeatable, reproducible and comparable. In other words, the quantity value is accurate and reliable. The consistency of measurement is not limited to the domestic circle but applies to the international circle as well. For example, the results of international comparison shall be consistent within the equivalent interval or protocol interval. The core of metrology is to confirm the measurement result and its validity and reliability, without which metrology will lose its social significance.

Traceability refers to the ability of any measurement result or value of a measurement standard to be linked to the measurement benchmark through a continuous comparison chain with specified uncertainty. This property makes all the same quantity values traceable to the source of measurement by calibration along this comparison chain to the same measurement benchmark, thus guaranteeing accuracy and consistency by technology. Otherwise, the value of multi-source or multi-head will inevitably cause technical and management confusion. Therefore, quantity traceability refers to the system of traceability from bottom to top through continuous calibration, while the quantity value is passed down from top to bottom to form a verification system.

Legality comes from the sociality of metrology because accuracy and reliability of quantity value depend not only on scientific and technical means but also on corresponding laws, regulations and administrative management. Particularly in areas which have a clear impact on the national economy and people's livelihood, which involve the public interest and sustainable development, and where special trust is required, the government must take the lead to establish the legal safeguard.

Otherwise, the accuracy, consistency and traceability of quantity value cannot be achieved nor can metrology play its due role. Forty-two countries such as the US, the UK and Germany have included metrology in their constitutions, as the basic requirement for the central authority and unified management of the country. The presidents of national metrological institutes in the US and Germany are appointed by the President.

In short, metrology is a kind of measurement derived from but more rigorous than measurement. It covers the whole field of measurement and guides, supervises and guarantees measurement in accordance with the provisions of the law. Scientific metrology provides measurement basis for engineering metrology and the development of new technology. It also provides technical safeguard for the legal system measurement; in other words, legal metrology is based on scientific metrology as the technical basis of administrative law enforcement. In fact, the development of science, technology, economy and society raises high requirements for the unity of units, accuracy and reliability of quantity value; the higher they are, the more important the metrology role. Therefore, metrology belongs rightfully to the national quality infrastructure enterprise.[5]

1.3 Measurement Error and Measurement Uncertainty

1.3.1 Measurement Error

It is difficult to trace the time when the concept of error originated, but as early as 1862, when Foucault used a rotating mirror to measure the speed of light on Earth, the result of the measurement was $c = (298{,}000 \pm 500)$ km/s, that is, at the same time as giving the measuring result, also gave the measures error. Thus, it can be seen that the concept of error emerged at least more than a hundred years ago. There are two difficulties in the application of error: problems in logic concept and problems in evaluation methods.[6]

Section 5.3 of "General metrological terms and definitions" in *The National Metrological Technical Code of People's Republic of China* (JF1001-2011) defines "Measurement error" as "the value of measured value minus reference value." In the annotation of definition, it explains that when a single reference value for a given conventional quantity exists, or the uncertainty of the measured value of the measuring standard calibration instrument can be ignored, the measurement errors are both known. When the reference value is true, the measurement error is unknown. A measurement error is a parameter that indicates whether the measured value deviates from the reference value, i.e. whether the result of the measurement is close to the referential value.

When the measurement error is known, the error is a positive or negative measurement value. The error is related to the measuring result, and the result can be obtained only through measurements. Therefore, the error can only be obtained by measurement and cannot be obtained by means of analysis and evaluation alone. For example, for the same measurement, when multiple measurements are carried out under repeatable conditions, different measurements may be obtained, the results may be different, and therefore the error of these different measuring results is different. When the measured result is greater than the true value, the error is positive. When the measured result is smaller than the true value, the error is negative. Therefore, the error cannot appear in "$\pm$" form. Measurement error is unknown because the true value of a quantity is the true size of the quantity itself when it is observed, and only a perfect measurement can get a true value. Since any measurement is defective and perfect measures do not exist, true value is an ideal concept. Since true value cannot be accurately obtained, the error cannot be accurately obtained either.

Let's use a diagram to describe more vividly the measurement results with error using the formula $M = T + e$ or $M = T - e$, where M is the measurement value, T is the true value and e is the absolute value of the error. Follow the arrow in the diagram (see Fig. 1.4); if the value e gradually increases, it indicates that different positions in the direction of the arrow represent different levels of the measurement result; the smaller the value of the e, the closer the measuring result M to the true value of T, indicating the higher the quality of the measured result M, and the higher level of measurements.

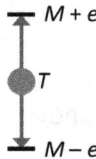

Fig. 1.4.   Diagram of measurement result with errors.

1.3.2 Measurement Uncertainty

The concept of measurement uncertainty was first introduced in 1963 by Eisenhart, a mathematical statistician at National Standards Board (NBS), the predecessor of National Institute of Standards and Technology (NIST), in the study of "the estimation of precision and accuracy of instrument calibration system." Once proposed, the concept of measurement uncertainty has attracted worldwide attention. The term "uncertainty" is derived from the English word "incertitude," meaning uncertainties, instabilities, doubts, etc., and is a qualitative term.

When the term is used to describe measurement results, the meaning is expanded to quantitative representation, i.e. the quantitative uncertainty of measurement result. Afterward, in 1986, international organizations unified and popularized the concrete expressions of uncertainty. The measurement uncertainty and its evaluation can not only be applied to the field of measurement but also to all other fields related to measuring. In 1986, seven international organizations, namely International Bureau of Weights and Measures (BIPM), International Electrotechnical Commission (IEC), International Organization for Standardization (ISO), International Organization of Legal Metrology (OIML), International Union of Pure and Applied Physics (IUPAP), International Union of Pure and Applied Chemistry (IUPAC) and International Federation of Clinical Chemistry (IFCC), draw up guidance documents on the evaluation of measurement uncertainty, under the requirement of International Committee of Weights and Measures (CIPM). In 1993, *Guide to the Expression of Uncertainty in Measurement* (GUM) and the second edition of *International Vocabulary of Basic and General Terms in Metrology* (VIM) were issued jointly by these seven international organization.

Subsequently, International Laboratory Accreditation Cooperation (ILAC) also expressed its recognition of GUM. These above-mentioned international organizations cover almost all areas related to measurement, indicating the authoritativeness of both GUM and VIM documents. In 1998 and 1999, China subsequently issued National Measurement Technical Regulations JJF1001 *General Metrological Terms and Definitions* and JFF1059 *Evaluation and*

Expression of Uncertainty in Measurement (JJF1059-1999), in accordance with the latest editions of GUM and VIM issued by the international organizations.

Measurement uncertainty is a parameter that quantifies the quality of the measurement result. In fact, due to the lack of comprehension and insufficiency of measurements, the measured value is decentralized, i.e. the result is not the same value each time but is scattered at a certain probability across many values within a given region.

Measurement uncertainty is generally composed of a number of components, some of which can be evaluated based on the statistical distribution of a series of measurement values as the measurement uncertainty of Type A characterized by standard deviation, while another component can be evaluated based on probability functions obtained from experience or other information as the measurement uncertainty of type B, or characterized using standard deviations. Usually, for a given set of information, the measurement uncertainty refers to the corresponding change of the given measured value. All of these components shall be understood as contributing to dispersion. From this, it can be seen that uncertainty represents an "interval," as opposed to error representing a "point," which is a parameter without a positive or negative number.

Let's also use a diagram to describe vividly the measurement results with uncertainty $M = C \pm \mu$, where M is the measured value, C is the central value of the measurement and μ is the measurement uncertainty. It can be seen from the diagram (see Fig. 1.5) that the true value of the measurements is unknown but with a certain probability falls within the range $[M - \mu, M + \mu]$. Following the arrows in the diagram, one can see that if the value of μ increases, it only

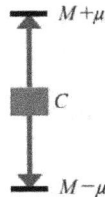

Fig. 1.5.   Diagram of measurement results with measurement uncertainty.

shows that the dispersion interval of the measured results increases, and it does not mean that the measurement value M is closer to the true value T. Since the true value is unknown, the level of the measurement value is actually the same for different positions in the direction of the arrow. This is also the essential difference between uncertainty and error in measurement results. The smaller the value of μ, the greater the probability that the measured result approaches the true value, and the higher the quality level of measurements.

The evaluation of measurement uncertainty is suitable for all kinds of measurement fields with various accuracy requirements. Therefore, it will permeate all fields of science and technology as shown in the following:

(1) **The establishment of national measurement benchmarks and measurement standards at all levels:** It applies to evaluate the measurement uncertainty of reproducible standard values upon the establishment of international metrological benchmarks and measurement of standards at all levels.

(2) **Comparison of measurement standards, testing equipment and measuring methods domestically and internationally:** It applies to the comparison domestically and internationally between measurement standards, testing equipment and measurement methods. Measurement uncertainty must be given when the participants report measurement results. By processing the data obtained by the participants, the consistency of the results is assessed. The comparison is a proof of the reliability of the result and a validation of the laboratory's technical capabilities.

Consistency evaluation of measurement results is obtained through the processing of data from reference laboratories. The comparison result is a proof of reliability of the result and a validation of laboratory's technical capabilities.

(3) **Value determination of reference materials (RMs) and release of standard reference data:** It applies to the release of value together with its uncertainty of reference materials (RMs) after the value of RM has been determined according to the prescribed method. The same applies to the release of standard reference data with its uncertainty.

(4) **Preparation of technical documents such as measurement method standards, inspection procedures and technical specifications:** The measurement uncertainty of the method should be analyzed and assessed in the preparation of measuring method criteria, technical specifications and verification procedures so that users can use them as a reference or as a fraction in evaluation of measurement uncertainties.

(5) **Measurement in the field of scientific and technical research and engineering:** Measurement uncertainty applies to all science and technology and engineering projects. All scientific and technological achievements must be evaluated in terms of measurement values with their uncertainties. For example, the analysis and estimation of measurement uncertainty is indispensable to the scheme demonstration of major projects; the measurement uncertainty shall be used correctly in the graduation thesis design of college students. Similarly, the knowledge of measurement uncertainty can be applied to measurement courses in colleges and universities.

(6) **Metrological authentication, metrological authentication, quality authentication and laboratory accreditation:** In the metrological authentication, metrological confirmation and quality authentication, reviews shall be carried out according to whether the measurement requirement, the measurement result and the measurement uncertainty can satisfy the requirement of use. In laboratory accreditation, the assessment of measurement range and uncertainty is an assessment of organization's technical competence.

(7) **Calibration and testing of measuring instruments:** Measuring instrument is an indispensable tool for people to measure. In order to ensure that its quality meets the requirements of use, it is necessary to carry out regular calibration or verification. At this point, the calibration value or the measurement uncertainty of the reference value is given.

(8) **Quality control and product inspection in production process:** It applies to the description of measurement results and evaluation of conformity on quality control and product inspection in the production process.

(9) **Trade settlement, healthcare, safety, environmental monitoring and quality control:** Where measurements are required, the results and the uncertainty of measurement shall be recorded for quality traceability. All goods or products must be inspected and qualified before they can be put on the market.

1.4 Certified Reference Materials

There are three forms of metrological references: metrological equipment reference, reference materials (RM) and reference measurement method. Reference materials (RM) is one of the expression forms of metrological references. It was created in 1906 when former American National Bureau of Standards (NBS) officially issued for the first time five RMs, including cast iron and converter steel. To date, RM has enjoyed a development history of over 100 years. It has been rapidly developed and applied all over the world, and has become a vital link in the measurement system. In 1980s, based on the needs and experience of major RM developers and user groups, ISO GUIDE30:1992 "Common terms and definitions for RM" defined RM jointly by seven international organizations including BIPM and ISO.

National Metrological Technical Code of China (JJF1005-1005) and "Common terms and definitions for RM" adopt this definition. In 2006, ISO GUIDE35:2006 "General principles and statistical principles for identification of RM" redefined RM. The definitions of RM and CRM are as follows:

(1) **Reference material (RM):** It has one or more specific characteristics that are sufficiently uniform and stable; the characteristics are established for their intended use in the measurement process.
(2) **Certified reference material (CRM):** It is a standard substance representing one or more of its specific characteristics by a valid metrological procedure. This CRM is accompanied by a certificate, which provides values and uncertainties for its specific characteristics, as well as a declaration of metrological traceability.

In 2007, joined by relevant international organizations, BIPM revised "International Vocabulary of Basic and General Terms in Metrology (VIM)" and published the third edition of VIM in 2007 (ISO/IEC Guide 90:2007), including RM and CRM. Their definitions are as follows:

(1) **Reference material (RM):** For certain characteristics that are sufficiently uniform and stable, the characteristics are established for their intended use in measurement and nominal characteristic testing.

(2) **Certified reference material (CRM):** It is accompanied by a document issued by an authority in which one or more characteristic values with uncertainty and traceability are obtained by reference to a valid procedure. There is no essential difference between the definitions given in the two ISO documents (ISO guide 35: 2006 and ISO guide 99: 2007); both are prepared and identified by reference to the valid procedures described in the same technical documents. The difference lies in the addition to the definition of "RM" in the third edition of VIM (ISO/IEC GUIDE 99:2007): (1) the intended use in nominal characteristics testing; (2) the "characteristics" of RM have two meanings: "Quantity value" and "Nominal characteristic value" and their respective uses. Values can be qualitative or quantitative. Qualitative CRMs provide nominal characteristic values. In the definition of CRM, ISO GUIDE 35: 2006 uses "Document" instead of "Certificate" and emphasizes that the main body to issue certificates shall be "the authoritative organization."

RM is characterized by stability, homogeneity and accuracy of the characteristic values, which are also the basic requirements of RM.

(1) **Stability:** Stability refers to the ability of a characteristic value of RM to remain within a specified range in the specified time and environmental conditions.

(2) **Homogeneity:** Homogeneity is the state in which one or more characteristics of a substance have the same composition or structure. In theory, if there is no difference in characteristic values between parts of a substance, then the substance is completely homogeneous for this given characteristic. However, whether there is any difference between characteristic values of each part of the substance must be determined by experiments.

Therefore, the so-called homogeneity means that the difference of characteristic values between each part of the substance cannot be detected by experiments. Hence, the concept of homogeneity includes characteristics of the substance itself and the measurement methods used, such as the precision of measurement method (standard deviation), the size of sample (sample size) and so on.

(3) **Accuracy:** Accuracy refers to the notion that the RM has accurate measurement or strictly defined reference value. Usually, in a RM certificate, both the reference value and its uncertainty are given. When the reference value is the conventional true value, the specification of metrological method when RM is used as "the calibration substant" is also given.

1.5　The Role of Metrology

1.5.1　The Role of Metrology in National Economy

BIPM states that national metrological system consists of the system of measurement units, national benchmark (basic standard) and value transfer system, metrological laws and regulations, national legal metrological institutions, metrological technical institutions and accreditation. Metrology is the basis of quality control, which is called the "eye" and "nerve" of industrial production. Metrology has entered our lives in many forms and in various fields, but most of time we do not feel its presence. In fact, the smooth operation of the metrological system is an important basis for social, economic and daily life.

1.5.2　The Role of Metrology in Basic Research

Metrology plays an important role not only in quality control and large-scale repeated and reliable manufacture of products but also in basic research. Dmitri Ivanovich Mendeleev once said, "No measurement, no science." The case of Nobel Prize in Physics for graphene is a good illustration of this. The structure of graphene has been simulated since 1940s, but the existence of the material has been difficult to determine because it has not been measured and proved experimentally. It wasn't until 2004 that physicists Andrea Garmr and Konstantin Novoselov of University of Manchester of UK succeeded in separating graphene from graphite by micromechanical stripping, and more importantly, that AFM and electrical measurements confirmed the existence of graphene, both scientists shared the 2010 Nobel Prize in Physics. In 2017, *Nature* published a paper "Metrology as the Key to the Replication of Results,"[3] written by experts at National Physical Laboratory (NPL) in the United Kingdom. The paper points out that scientists in all disciplines must work with measurement specialists to ensure comparable results. An example of radiation therapy is given that radiation therapy is the practice of using ionizing radiation to kill or affect cancer cells. Although there are strict regulations on the amount of treatment a patient can receive in a clinical setting, basic research laboratories that study the effects of radiation on cells do not have similar regulations. However, in 2013, a research report by NIST, USA found that only 7% of valuable annual articles published in the journal *Radiation Research* cited established dose standards and guidelines. NIST concluded that radiobiometry "is frequently inadequate, thereby weakening the reliability and repeatability of the results," which has led to barriers to the translation of preclinical research into clinical practice and unnecessarily increased the number of animals used in the study. At present, many research centers in the US are working on dose standardization, which illustrates the important role accurate measurements (metrology) play in the repeatability of the results of basic research and the industrial application of basic research results. The paper emphasizes that metrologists represent time-saving and accuracy improvements in measurement results. In many cases, breaking through the bottleneck of technical progress depends on the solution of key measurement technology and the reliability and repeatability of research results.

1.6   Overview of Materials Metrology

Materials science is the science that studies the structure, properties, production process and performance of materials, as well as their interrelationships. Materials science is the result of the intersection and integration of multiple disciplines, an applied science indispensable to engineering technology. High-quality R&D of materials and production are inseparable from technology, equipment and production. Equipment is not only the carrier of technology and materials, but it also provides service to the realization of materials and processes, thus these aspects are forming a triangle. Only when closed loops are formed can high-quality R&D and production of materials be meaningful; nothing can be achieved without any of them. An important objective of the materials industry is to make materials into substances, and substances into equipment, and equipment to good use.

Materials metrology is a subject concerning the measurement and application of materials and their production processes. It is an activity aiming to study the unity of measurement units and the accuracy and reliability of measurement values within the system of materials research and development, production and manufacturing. As a new metrology field, materials metrology is the inheritance and continuation of traditional measurement. Traditional metrology is aimed at accurate measurement of single parameters, such as length, heat, force, electricity, light, magnetism, sound and chemical composition. Traditional metrology focuses on the reproduction, traceability and transfer of SI International Unit of an instrument, which realizes the value of a single parameter. The measuring object of material metrology is changed from the measuring equipment to both technologies of the value traceability and value transfer of material and its production process.

When the material itself is the object of measurement, material metrology means complex technology, including measurement of multiple parameters of material structure, composition and performance, and conducts production assessment based on comprehensive analysis according to the above measurement results. When the material production process is the object of measurement, material metrology means the layout of key parameters from batch consistency and product quality and multi-dimensional accurate measurements of material measurands and their applications.

Materials metrology is realized by studying the reduction of measurement uncertainty in the measurement process.[7] Uncertainty is defined as the degree of dispersion reasonably assigned to the values to be measured, and it is a parameter associated with the measurement results.[5,6,8] In layman's terms, uncertainty is the range that fluctuates around the central value of the measured values by different laboratories or different operators with a certain probability. For example, the X-ray diffraction angle of a material $2\theta = 23° \pm 2.0°$, meanwhile, $2.0°$ is an uncertainty. We can imagine that the larger the range (that is, the greater the uncertainty), the greater the dispersion of measured values and worse the consistency. At this time according to these measurement values with very large uncertainty, on the one hand, from the perspective of scientific research, the trial-and-error cycle of R&D will be prolonged by adjusting the experimental program, resulting in a longer cycle of R&D of new products. On the other hand, in terms of product quality, it will cause the worsening consistency of product quality. Therefore, reducing uncertainty is an important means to shorten the R&D cycle and improve product quality. In order to reduce the uncertainty as much as possible, it is necessary to analyze the sources of uncertainty and find the biggest sub-uncertainty component to the combined uncertainty. This component is also the most important factor affecting the accuracy of the measurement results. Once the most influential factor (the biggest sub-uncertainty source) is identified, targeted and purposeful improvements can be made in the production and scientific research process. Such purposeful experimental research will help greatly shorten the R&D cycle and quickly solve the difficulties in product quality improvement. Therefore, materials metrology is one of the quality infrastructures supporting the development of materials industry.[9] The goal of material metrology is to support the materials industry to achieve the objective of making materials into substances, and substances into equipment, and equipment to good use.

1.6.1   Research Contents of Materials Metrology

There is a big difference between materials metrology and traditional metrology. Traditional metrology takes reproduction, traceability and transfer based on a single SI basic unit of value (such as meters in length, seconds in time and Amperes in electricity) as its research content and the equipment to achieve the goal of accurate values as its research object (Fig. 1.6(a)). In contrast, the

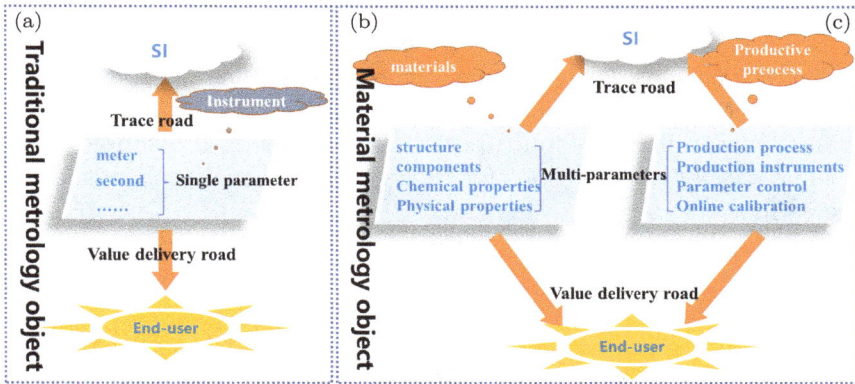

Fig. 1.6. Schematic diagram of research contents of traditional metrology and materials metrology: (a) research content of traditional metrology; (b) research content of the metrological technology of the material itself; (c) research content of metrological technology in material production process.

research object of materials metrology takes different and various materials, with their measuring equipment and production process as its research content, including key parameters involved in materials development and quality control as well as parameters involved in production process control. For example, when describing a material completely and accurately, its microstructure, composition, chemical and physical properties are key metrological parameters (Fig. 1.6(b)), while the accurate measurement of these parameters is closely related to measuring equipment and methods. Therefore, metrological technology research content of the material itself (R&D and quality control) includes establishment of measurement equipment traceability, measurement method standardization, development of CRMs needed for value transfer and method validation, and an international comparison to make the measurements internationally equivalent.

Research on metrological technology of material production process includes three-dimensional layout control and online calibration of various parameters in manufacturing process flow (Fig. 1.6(c)). The progress of science and technology has placed higher requirements on the R&D of new materials. Among them, reducing R&D cycle and improving R&D efficiency have become an important subject of advanced material R&D. Therefore, the content of material measurement research is further extended. Building materials database based on accurate measurement of material structure and properties has become a new research content of materials metrology in national metrological institutes of various countries. Thus, material metrology is a multi-parameter metrological technology.

The specific explanations for each element of materials metrology are as follows:

(1) The principle of measurement usually refers to basic laws with universal significance, which is the theoretical basis of materials metrology, guiding the whole process of materials metrology according to scientific principle.

(2) Measuring equipment usually refers to the equipment which is based on measuring principles and meets the needs of precise measurement of material characteristic parameters and the requirements of uninterrupted traceability to basic units of SI.

(3) The measurement method usually includes a pretreatment method to obtain the sample to be measured which meets the measurement requirement and the optimized measuring conditions and operating procedures to ensure the consistency of measurement results.

(4) Data processing refers to the process of iteration, fitting and peak splitting on measurement results which are not displayed directly according to the measurement principle.

(5) The evaluation of uncertainty refers to the process of calculating and identifying all uncertainty sources from measuring equipment, measuring method and data processing according to the principle of measurement.

(6) RM of materials metrology refers to the RM that is used for calibration of measuring equipment for material characteristic parameters and ensures the validity and consistency of measurement methods for specific material parameters.

(7) International (metrological) comparison means the process of comparison, analysis and evaluation between the amounts of values reproduced or maintained by the same measurement benchmarks and measurement standards on the platform of international organizations under specified conditions.[9]

(8) The material measurement elements in the process of material production and manufacture refer to the layout control and online measurements of various parameters (such as temperature, humidity, flow rate and thickness) in the material production process and calibrations of their measuring equipment.

(9) Materials database is the basic parameter of the relationship between material structure and its performance, which is the foundation of advanced material design and simulation research. Based on internationally recognized accurate data, desired advance material design and simulation solution can achieved, which may guide advanced material preparation experiment, thereby shortening R&D cycle of new materials.

1.6.2 Research Outcomes and Social Services of Materials Metrology

The main outcomes of materials metrology are standard equipment, CRMs (including reference data) and standard methods (including calibration specifications). The means of service is the use of CRMs, standard methods, standard devices, databases and so on by which social services are provided through calibration and testing, in order to achieve confidence in data and assurance of product quality in the development of materials and manufacturing process. This confidence comes from the calibrated equivalent and consistent results of various international metrological comparisons by different national metrological institutes (NMIs) on the platform of international metrological organizations. Figure 1.7 shows the metrological traceability involved in materials research and production practice as well as routes of social service. Metrologists mainly follow the path shown in Fig. 1.7(a) to conduct research on measuring equipment, measurement method and data processing analysis for specific parameters of a given material according to the principle of measurement. On one hand, based on the study of measurement principal and measurement equipment traceability technology as well as RM research, the measurement value is traceable to SI basic unit to ensure the accuracy and reliability of the measurement results of the measuring equipment. On the other hand, on the basis of

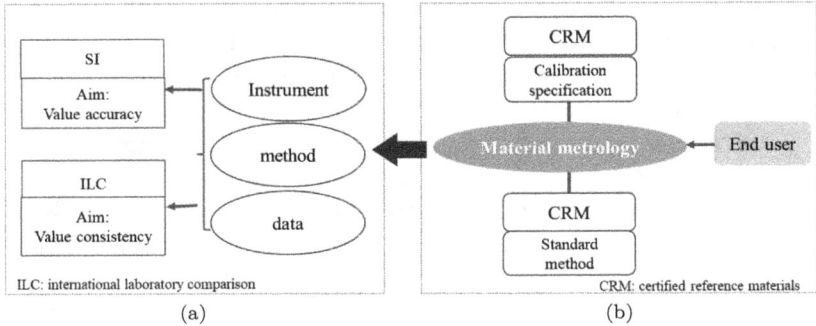

Fig. 1.7.   Illustration of accurate and reliable paths in materials research and production practice: (a) research path of metrological researchers; (b) traceability path of end users.

equipment traceability, the consistency of measurement method and RM research is carried out through international comparison to ensure the consistency and equivalence of measurement results and develop reference methods.

The end-user traceability path is shown in Fig. 1.7(b). Based on available CRMs, calibration specifications and reference methods, measuring equipment and methods in the working environment are traced to SI basic unit. The equipment is calibrated and measured by a combination of calibration specifications and CRMs. The sample to be measured is measured by a combination of standard methods and CRMs. Finally, the accurate measurement value of the calibrated equipment is transferred to the end user, ensuring end-user measurement results are consistent and equivalent worldwide/nationwide. In this way, quality assurance of materials industry and the promotion of basic support is completed, thus achieving the important goal of turning materials into substances and good substances into handy equipment.

1.6.3   Current Situation of Materials Metrology Worldwide

Materials metrology is the combination of scientific metrology and industrial metrology. Materials metrology research of BIPM can be traced back to the studies on the structure, thermal and mechanical properties of fern alloys in 1890s. Researcher Charles

Édouard Guillaume won the Nobel Prize in Physics in 1920. At present, Surface Analysis Working Group (SAWG),[10] set up by BIPM within the framework of Consultative Committee on Quantity Material (CCQM), is developing international comparisons on film thickness, surface composition and Raman spectra. Versailles Project on Advanced Materials and Standards (VAMAS)[11] is carrying out international pilot studies on comparative and collaborative research of morphology, composition, purity/impurity content and mechanical/thermo-dynamic/electrical/optical properties of the structure of carbon-based nanomaterials, thin film materials, semiconductor materials and polymer materials. The materials metrology technology committee (TCMM) of Asia Pacific Metrology Programme (APMP)[12] is carrying out international laboratory comparisons of film thickness, mechanical properties and crystal structure.

The earliest institutes of metrology in the world that carry out metrology research and provide the most developed social service on metrology are from USA, UK and Germany. These three national metrological institutes are currently the most advanced metrological institutes in the world; their metrological thoughts and metrological activities play a leading role in the world.

NIST[13] is America's NMI; one of the missions given at the beginning of its existence in 1901 was the measurement of materials, namely the measurement of physical constants and material properties, and was adopted by Congress. Today's front page of the NIST website highlights NIST's core philosophy: measuring, innovating, leading and working with our colleagues in industry and science to stimulate innovation and improve life quality. NIST has a material measurement science laboratory, whose main task is to carry out basic and applied research on the composition, structure and performance of industrial materials and production process, to ensure measurement quality through the development and delivery of CRMs, reference measurement procedures, assessed data and operational guidelines. NPL,[14] UK's National Physical Laboratory, says on the front page of its website that NPL's mission is to provide world leadership solutions for business and government, focus on R&D and innovation, improving quality of life and promoting trade. German Federal Institute of Physics and Technology (Die Physikalisch-Technische Bundesanstalt (PTB)) and German

Federal Institute of Materials Research and Measurement (Bundesanstalt für Materialforschung und -prüfung (BAM)) are two national metrological institutes of Germany. The former tends toward the measurement of physical parameters while the latter toward the measurement of chemical and material parameters. BAM describes its role and responsibility on its website in this way[15]: BAM represents the high standards of chemical and technical safety in Germany and the global market and is the technological foundation for further upgrading the quintessence of "Made in Germany" quality culture. It conducts physical and chemical testing and evaluation of materials and devices and advises the federal government, industry and national and international organizations in the materials field.

Take the research of NIST as an example. On the website of Material Measurement Laboratory (MML) attached to NIST are written clearly the missions of material measurement:[16] to assure and enhance the accuracy and reliability of measurement science. "The laboratory supports the NIST mission by serving as the national reference laboratory for measurements in the chemical, biological and material sciences. Our activities range from fundamental and applied research on the composition, structure and properties of industrial, biological and environmental materials and processes, to the development and dissemination of tools including reference measurement procedures, CRMs, critically evaluated data, and best practice guides that help assure measurement quality" (NIST website). In developed countries, the concept of metrology is deeply rooted in people's minds. Both basic research and industry have a deep understanding of the importance of measurement. Enterprises will take the initiative to seek support from NMIs for technical services like enterprise product upgrading and quality control. Equipment manufacturing enterprises in developed countries often cooperate with their NMI to carry out R&D, production and sales of all kinds of material characterization and measurement equipment worldwide. VLSI Corporation, in cooperation with NIST, worldwide sells standard RMs, whose values are traced to NIST standards.

In China, material measurement, as a new field, is not popular in the society. While material production enterprises know something about verification and calibration of traditional measurement; their knowledge of the specific help of material measurement for enterprise product quality improvement and control is not enough. But universities and research institutes with basic research as the main body have no knowledge basically of material measurement except material characterization because there is no market trade. This has exposed some problems, such as rejection of research papers due to unreliable data. In scientific research institutes materials researchers tend to think that no material measurement is needed since they are working on characterizations and measurements, which makes it difficult to popularize the concept of material measurement. However, the control of accurate measurement results in materials metrology can greatly shorten the trial-and-error cycle for basic research and industrial product development and improve R&D efficiency and product quality. China has a very typical phenomenon, that is, the more developed economic areas have a stronger need for understanding and demand for material measurement, while the less developed economic areas have hardly any material measurement. Similar phenomenon is observed in overseas countries. The more developed the country/area, the more it needs material measurement, suggesting the supporting role of material measurement in economic development.

National Institute of Metrology, China (NIM, China) set up an advanced materials metrology laboratory in 2012 and has started research work on measurement technology of carbon-based nanomaterials, thin film and particles since "The 12th Five-Year Plan" and participated in the comparison activities of international metrology organizations. The measurement technology research on the material crystal structure, geometric structure, chemical structure, thermo, electric and optical properties and other parameters are carried out, and the value traceability and delivery path for glancing incidence X-ray reflection technique, X-ray diffractometer, transmission electron microscope, Raman Spectroscopy, Seebeck coefficient of bulk materials, etc. were established. Moreover, more than 20 kinds of CRMs are supplied to industry, such as nanoscale film thickness CRM (used for calibration of thickness, calibration of middle or low magnification of transmission electron microscope), gold {111} interplanar spacing CRM (used for calibration of high magnification of transmission electron microscope), Raman shift and Raman relative intensity CRM (used for calibration of Raman spectra), and single atomic step height CRM (used for calibration of AFM in smaller than 1 nm scale).[17] Furthermore, this laboratory led and participated in more than 10 international comparisons under CCQM/SAWG, APMP/TCMM and VAMAS/TWA41.

Laboratory of Advanced Material Measurement (LAMM) of NIM, China, plays an important role in international metrology and standard organizations. Dr. Ren Lingling, the leader of the LAMM, was the former chairman of APMP/TCMM, co-chairman of VAMAS/TWA41 and China registered representative of ISO/TC229, and Dr. Yao Yaxuan is an registered representative of CCQM/SAWG. The lab uses these international organization platforms to lead and participate in international comparisons, establish ISO and IEC international standards based on these comparison results and promote the development of technologies in this field by sharing of these research results.

1.7   International Organizations of Metrology

As can be seen from Section 1.6.3, materials metrology is closely related to the activities of international organizations of metrology. This is determined by the characteristic requirements of accuracy, reliability and traceability of metrology. It is an important technical path to carry out international metrological comparisons in globally recognized international organizations, which means meeting requirements of these characteristics (accuracy, reliability and traceability of values). So, here is a brief introduction to international organizations concerned with materials metrology. Currently, the international organizations related to materials metrology in the world are BIPM, APMP and VAMAS.

1.7.1   Bureau International des Poids et Mesures (BIPM)

The idea of a unified international system of measurement first appeared at the first World's Fair in London, UK, in 1851. At the fair, depending on the countries of origin, the technical specifications of the exhibits ranged from English and metric systems to other units, which makes the selection of many award-winning products extremely complicated, hence the idea of a unified international system of measurement. At the Paris World's Fair in France and the Second Congress of Statistics in 1855, the initiative was put forward for the establishment of the International Committee for a unified system of measurement. On May 20, 1875, 17 countries signed the Metre Convention and set up BIPM. Then, the manufactures of meter and kilogram prototypes were carried out to meet the needs of science, technology, manufacturing and commercial activities.

BIPM is the executive body of General Conference on Weights and Measures and CIPM, and it is a permanent world research center for metrology. BIPM is a neutral and autonomous body that is not affiliated with any existing intergovernmental organization and does not join any international alliances or associations, only with United Nations Educational, Scientific and Cultural Organization (UNESCO), International Atomic Energy Agency (IAEA), European Atomic Energy Community (EAEC), OIML, etc. It enjoys extrajudicial rights and immunities in the territory of France in accordance with the agreement between BIPM and the French government, and the government recognizes it as a public-benefit institution. Its primary tasks are to ensure the uniformity of measurements worldwide, with specific responsibility for establishing benchmarks for major measurement units, preserving international instruments, organizing comparisons between national and international benches, coordinating the measuring work on basic physical constants and coordinating related measuring techniques.

The General Conference on Weights and Measures is the highest form of organizational form of the Metre Convention and is convened every four years if there are no exceptional circumstances. CIPM, the permanent organization of the Metre Convention, meets once a year in Paris from September to October. Contributions to the convention are made by each Member State, all of which are spent on research funding for the International Bureau of Measurements, salaries for staff and operating expenses.

The academic body under CIPM, Advisory Committee, which is responsible for the study and coordination of the academic issues of measurement in its specialized field, has established 10 Consultative Committees in Electricity and Magnetism (CCEM), Thermometry

(CCT), Time and Frequency (CCTF), Length (CCL), Ionizing Radiation (CCRI), Units (CCU), Photometry and Radiometry (CCPR), Mass and Related Quantities (CCM), Amount of Substances in Chemistry and Biology (CCQM) and Acoustics, Ultrasound and Vibration (CCAUV). May 20 is "World Metrology Day."

The mandate of BIPM is mainly to develop and maintain international unit benchmarks. Early emphasis was placed on the calibration of international unit benchmarks for member states, but in recent decades, more emphasis has been placed on the benchmark research of basic units.

In October 1999, the 21st General Conference on Weights and Measures passed a resolution on "Calibration and Certification of Mutual Recognition Arrangement issued by National Measurement Bases Standards and National Measuring Institutions" (MRA) and organized critical quantitative comparison at the international level within BIPM, with BIPM playing a guiding and core role. All key units' comparison data are entered into the BIPM database (KCDB) and are available on the Internet. It can be seen that there will be a considerable reduction in the work of BIPM to carry out inspections directly in individual countries in the future.

Here is more introduction to Mutual Recognition Arrangement (MRA). MRA can be regarded as a mechanism for establishing the accuracy and reliability of the calibration and measurement capability certificates of the national metrological institutes, and for building customer confidence in the national metrological institutes. The objectives of the MRA are to establish equivalence between national metrics standardized by National Metrological Institute (NMI), to mutual recognition between calibrations issued by National metrological institutes and certificates and thus to provide a reliable technical basis for governments and other sectors to conclude broader international agreements on international trade, commerce and management. Mechanisms for achieving these objectives are as follows: international matching of measurement results, known as key comparison; auxiliary matching for measuring results; and the quality system and proof of capacity of NMIs. This seemingly simple mutual recognition mechanism, proposed from the 1986 draft to the final agreement reached in 1999, reflects that the implementation of the international comparisons is itself something that requires strict regulation, and published "Guide for key comparison."[18] In the last two decades since the adoption of MRA, the work undertaken within the framework of the Metrology Convention has been much richer than before, and the relationship between NMIs has become closer. Comparative research on results, as well as multi-faceted contacts required for the implementation of a mutually recognizable quality system, have brought researchers in different countries closer together. This reflects the urgent need for achieving international equivalence of verifiable measurement results, as well as the wider consensus that reliable measures provide fundamental support for many important government decisions.[4] Comparative research on results, as well as the multi-faceted contacts required for the implementation of a mutually recognizable quality system, has brought researchers in different countries closer together.

For emerging material metrology (measurement), an important way to international recognition is international comparison. Only through strict international comparisons, the results of measurements are internationally consistent and equivalent, can the established measuring capacity be proven to be reliable and effective.

References

[1] Zhu, Xing, Benjamin Skuse. New definition of Basic Units in International International System [J]. *Physics*, 2018, 47(12): 795–797.

[2] Division of Metrology, State Bureau of Quality Technical Supervision. *General Metrological Terms and Definitions* [M]. Beijing: China Metrology Press, 2001.

[3] Sene M, Gilmore, I. Janssen, J.T. Metrology is key to reproducing results [J]. *Nature*, 2017, 547(7664): 397–399.

[4] Quinn, Terry. *From physical object to atom: Exploration of International Bureau of Weights and Measures and ultimate standard of measurement* [M]. Zhang, Yukuan (trans). Beijing: Quality Inspection of China Press, 2015.

[5] Metrology Department of the State Bureau of Quality and Technical Supervision. *General Terms in Metrology and Their Definitions* [M]. 3rd edition. Beijing: China Metrology Press, 2001.

[6] Ni, Yucai. *Evaluation of Uncertainty in Practical Measurement* [M]. 3rd edition. Beijing: China Metrology Press, 2009.

[7] Ren, Lingling. Implementation experience of the NQI technology whole chain in Graphene material [J]. *Journal of Metrology*, 2019, 40(3): 538–540.

[8] JJF1059.1–2012. *Evaluation and Expression of Uncertainty in Measurement* [S].

[9] JJF 1117–2010. *Measurement Comparison* [S].

[10] https://www.bipm.org/en/about-us/ [OL].

[11] https://www.vamas.org/ [OL].

[12] http://www.apmpweb.org/fms/general.php?tc_id=MM [OL].

[13] https://www.nist.gov/ [OL].

[14] https://www.npl.co.uk/ [OL].

[15] https://www.bam.de/Navigation/EN/Home/home.html [OL].

[16] https://www.nist.gov/mml [OL].

[17] https://www.ncrm.org.cn [OL].

[18] Guideline for Key Comparisons, March 1999, was replaced in 2010 by Measurement Comparisons in the context of the CIPM MRA, CIPM MRA-D-05.

Chapter 2

National Quality Infrastructure and Metrology (Measurement) of Graphene and Related 2D Materials

Lingling Ren

When it comes to the graphene industry, we have to talk about quality, as it is a matter of great concern to people. Since 2014, the central government of China has clearly proposed that the focus of development should be shifted to improving quality and efficiency. "Promote the transformation from made in China to created in China, the transformation from China speed to China quality, and the transformation from Chinese products to Chinese brands." Moreover, Chinese government in 2019 highlighted the importance of quality, calling for vigorous enhancement of the quality and efficiency of development. It put forward such major propositions as insisting on quality first, promoting quality reform, enhancing quality advantage, building a powerful country with high quality and realizing high-quality development.

China's national standard GB/T 190-2016/ISO 90: 2015 *Quality Management System: Fundamentals and Terminology* defines quality as "the degree to which a set of inherent characteristics satisfies a requirement." Characteristics include the inherent and conferring characteristics of some event or object. An inherent characteristic is a built-in feature of some event or object, such as the diameter of the screw and the material optical, electrical and thermal properties. A conferring characteristic is not a built-in feature of some event or object but the added features of the product due to different requirements upon completion of the product, such as product price, hardware product transfer time and shipping requirements (e.g. mode of transport) and after-sales service requirements (e.g. warranty time). As a result, quality formation process needs to be considered from value chain (conferring characteristic) analysis and characteristic parameter (inherent characteristic) measurement. Conferring characteristics depend more on management system, while inherent characteristics can be presented and quantitatively evaluated by objective testing. Therefore, the quality concept described later refers to the quality concept of inherent characteristics of the product.

2.1 Quality Concept in Ancient China: Weights and Measures

In fact, the concept of quality control has been appeared around since the time of Fuxi and Nüwa, founders of Chinese civilization, as attested by the totem of Fuxi and Nüwa (Fig. 2.1). It is evident from the totem representing prosperity that Nüwa holds *gui* in the right hand while Fuxi holds *ju* in the left hand, which shows that *gui* and *ju* in the totem are indispensable to prosperity. The *gui* and *ju* here stand for measurement. At present, this totem of Fuxi and Nüwa, the earliest symbol of measurement in China, is copied in the hall of Industrial Technology Research Institute (ITRI) in Taiwan, China.

Xunzi, a great thinker and politician in ancient China (circa 313–238 BC), clearly expounded the concept of quality of ancient China in his *Xunzi Theory of Rites*: "Once the rules are set, the standards cannot be broken." That is, standards (rules) are formed through the measurement (*gui* and *ju*) to regulate the quality. The words of wisdom people are more familiar with come from *Mencius Li Lou Shang*: "Nothing can be accomplished without *gui* and *ju*." It further expresses the concept of quality (making square and round objects) based on measurement (*gui* and *ju*). This concept of quality has run through the whole history of Chinese civilization since the beginning of its formation. In the primitive society, people had adopted the human hand as the equipment for volume measurement. They

Fig. 2.1.   Totem of Fuxi and Nüwa.

called the amount held by one hand as *yi* and the amount held by two
hands as *shen*. This provided a capacity standard for the exchange of
goods which made commodity transactions traceable. But this capacity standard deviation is very big; its main uncertainty sources are
from the following: First, human hands vary in size; second, the ways
and conditions of how to measure the "hand to hand" are held vary.
Although this measurement method with large uncertainty enabled
the commodity trading in life to follow a certain standard, it was far
from enough for large-scale industrial production and manufacturing.
To promote the development of the state, Shang Yang of the state
of Qin supervised in person the production of *Fang Sheng* (a square

bronze measuring instrument), which became the standard measuring equipment in Qin during the Warring States period. This *Shang Yang Fang Sheng* was the standard gauge of Qin Dynasty issued which held the highest rank "Great manufacture" in Qin Dynasty (now, it is collected in the Shanghai Museum). This bronze ware was cast in the Warring States period; it was 2.32 cm in height, 18.7 cm in total length, 12.4 cm in the length of the inner port, 6.9 cm in width, 2.3 cm in depth and 202.15 ml in volume. The production of *Shang Yang Fang Sheng* allows for accurate measurements of capacity indirectly through objective length, thus largely reducing the uncertainty (error) of capacity measurement. This is an indication that as early as 300 BC, our ancestors had realized that capacity was not a basic unit and had derived volume from length using the scientific method of "measuring volume in length," which was both accurate and convenient. Unified rules of measurement had laid down unified standards, necessary for the stability of the economic order, provided convenient conditions for the stability of economic order, facilitated conditions for people to engage in economic and cultural exchanges, and promoted economic, cultural exchange and development. The Qin government had a standard of measurement, promoted the great development of Qin industry and commerce, and provided solid material conditions for Qin to become the seven majesty of the kingdom of war, until the unification of the country. As the historical witness of Shang Yang's reform to promote economic development, *Shang Yang Fang Sheng* is the embodiment of China's ancient concept of quality with measurement as the basis. Qin Shi Huang achieved great power through Shang Yang's reform and finally unified the other six states; this process shows that China's ancient concept of quality was unified Weights and Measures, which could promote the high-quality development of national economy.

2.2 Modern Quality Concept: National Quality Infrastructure

Entering the modern era of high-speed industrial development, national quality infrastructure (NQI) has become the modern quality concept. NQI is a generic term for the quality institutional framework necessary for a country to establish and implement measurement,[1] standards[2] and qualification evaluation (usually referring to the collection of inspection, testing and certification),[3] including legal system, management system, technical system and so on. As with infrastructure such as transportation, communications, water supply, cultural education, medical and healthcare, NQI also serves as the foundation to safeguard economic and social development. The idea of NQI first appeared in the report "Innovation in Export Strategies" in 2005 jointly proposed by United Nations Conference on Trade and Development (UNCTAD) and World Trade Organization (WTO). In 2006, United Nations Industrial Development Organization (UNIDO) and International Organization for Standardization (ISO) issued a study that officially introduced the concept of NQI, defining "metrology, standardization, conformity assessment" as the three main components of NQI, which has become the three pillars of sustainable development of the world economy in the future, and are important technical means of improving productivity, inning life health, protecting consumer rights, protection of the environment, safeguarding safety and improving quality. These three pillars can effectively support social welfare, international trade and sustainable development. Since it was launched, the concept of NQI has aroused great attention of the international community and has been widely accepted by the international community. Metrology, standardization and conformity assessment (inspection, testing, certification, and accreditation) play an indispensable and fundamental role in the progress of human society and industrial development.[4,5] Metrology in the three components of NQI refers to the activity of achieving uniformity of units, ensuring accurate and reliable measurement values, and is the science and application of measurement. Standards are a normative document that is commonly used and reused in order to obtain the best order in a given range, consensus and approval by recognized bodies in order to obtain the best order within a certain range. Inspection and testing in conformity assessment are activities conducted when necessary for conformity assessment to analyze and measure, inspect and test the product safety, performance and other characteristics or parameters. Certification is a third-party testimony of products, processes, systems, and personnel, similar to the "guarantors" and "witnesses." Accreditation is the qualification

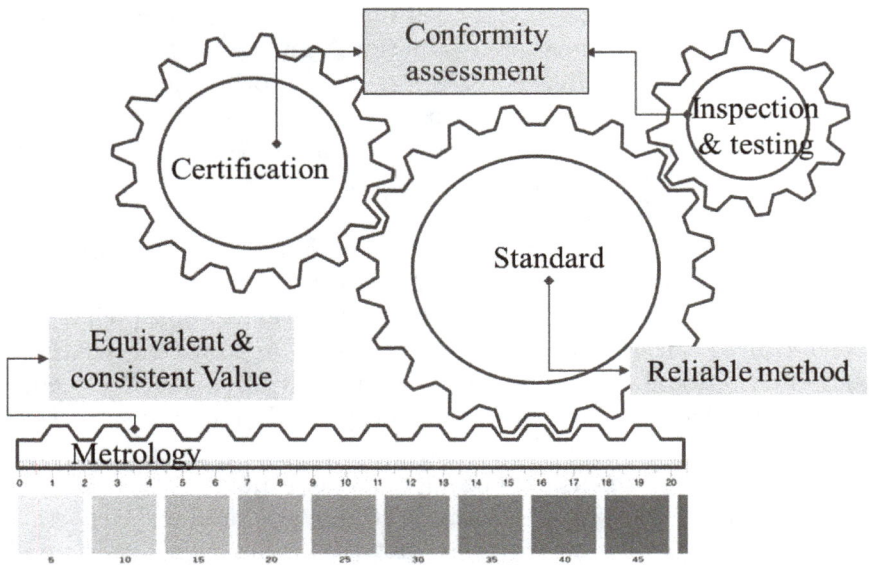

Fig. 2.2.   Role of the various components of NQI in the quality control process.

examination of inspection, certification and other institutions. As a complete technical chain, the various components of NQI are like gears that have to be closely interlocked to function, as shown in Fig. 2.2. Metrology as a ground base, which solidly supports the formulation and implementation of high standards, is the basis of quality control; standards, based on the results of metrological technology, are developed by consensus as e-normative documents, becoming the basis for certification approval and inspection testing. Therefore, the standard is the carrier of measurement value, whether inspection, testing and certification comply with the standard to determine whether the standard is met. To put it simply, metrology solves the problem of accurate measurement; the criteria depends on the accuracy in need; then, it is necessary to determine how the standard is implemented through certification approval and inspection testing. Therefore, metrology is the basis for quality control, standard is the premise for leading quality improvement, and conformity assessment is the means of transferring trust in quality.

NQI contributes significantly to economic growth. According to related research, the contributions of standardization in China,

Germany, France, Britain and Austria to their own economic growth have reached 7.88%, 27%, 23%, 12% and 25%, respectively. The measurement activities in industrialized countries have contributed 4–6% of GDP. Over 80% of the trade must be realized through measuring. Therefore, only on the solid foundation of metrology can high-quality standards be established, and inspection and testing be conducted according to high-quality standards to obtain high-quality data. Only from high-quality data can high-level certification be provided for product quality, thus promoting product quality and truly realizing the goal of a quality strong country. The integration of NQI elements for the entire chain implementation depends on a consistent core: consistency of key parameters and accurate measurement values. For example, reduced Graphene Oxide (rGO) product, the most widely produced graphene related two-dimensional materials on the market today, is a material obtained by reduction of graphene oxide, which is made up of a single layer of graphene as a structural unit and composed of up to 10 layers, as defined by national standards.[6] In contrast, the indicator requirements of customers and enterprises raised by businesses in the product trade are rGO in complete crystal form within five layers. Based on this requirement, NQI research begins with the research of accurate measurement method, as a result, to establish effective methods to measure the number of layers (using atomic force microscopy, Raman spectroscopy and transmission electron microscopy) and the crystal structure (using X-rays diffractometer). Based on reliable measurement methods, for product quality improvement, enterprises have confidence in the data obtained on the basis of these measurement methods to accurately judge the quality of products. For R&D, in order to raise competitiveness, enterprises try to improve product quality by improving the production process according to reliable measurement data, thus ensuring a competitive position in the market. It is evident that for both scientific research and industry, accuracy, reliability and equivalence of measurement value are the fundamental guarantees for improving research efficiency and product quality.

Quality improvement is achieved by reducing measurement "uncertainty." The definition of uncertainty is a parameter that characterizes the reasonable discrepancy of the measured value and is associated with the result of the measurement.[6–8] Typically, uncertainty is the range of values measured by different laboratories or different operators around the center of measurement values at a certain probability, such as 2.0° of $2\theta = 23° \pm 2.0°$. We can imagine that the larger the range of fluctuations (i.e. the greater the uncertainty), the more dispersive the measurement values, and the worse the consistency. Based on the large measurement uncertainties, on the one hand, from a scientific research perspective, adjustment of experimental program development trial cycle will be prolonged, resulting in longer new production development cycle; on the other hand, from the view of product quality, the product quality coherence will be worse. Therefore, the reduction of uncertainty is an important means of shortening R&D cycle and improving product quality. This is also one of the important contributions of measuring technology to quality improvement and assurance.

The establishment of an accurate and reliable measurement method needs the research of metrological technology, including traceability of measurement results and international comparison. The four major international organizations, namely International Bureau of Weights and Measures (BIPM), International Organization of Legal Metrology (OIML), International Laboratory Accreditation Cooperation (ILAC) and International Standardization Organization (ISO), signed jointly *The Joint Declaration on Metrological Traceability by BIPM, OIML, ILAC, and ISO* on November 9, 2011, which focuses on the following: metrological traceability is one of the key factors for the measurement result to enjoy worldwide equivalence and build trust. On the basis of traceability research, only by international comparison[9] can the measurement result be equivalent and consistent, and the measurement method accurate and reliable. International standards, national standards, group standards and other standards have to be established based on accurate and reliable measurement methods so that testing can be conducted on these standards and measurement results can be comparable across different labs (whether international or domestic), bringing reliable confidence to industry and market. It is clear that the core of NQI is the accuracy, reliability, equivalence and consistency of data with metrological technology as the source. In terms of the product, its quality is determined by the quantity value of the parameter, while the data of the quantity value of the parameter comes from the measurement technology. In essence, each technical research of NQI is a basic research based on data, among which reducing uncertainty is an important means to achieve quality assurance and promotion.

2.3 Basic Position of Metrology in National Quality Infrastructure

NQI starts with metrology. The reason for large industrial production exceeding and replacing a small-scale agricultural economy is that advanced metrological technology can drive quality improvement of the entire industrial chain from basic materials, basic components to major equipment, key processes and even the final product to provide a road map for quality improvement, thereby optimizing the industrial structure and increasing the added value of products and services. Metrology is the basis of quality control, known as the "eye" and "nerve" of industrial production. The quality control level of each link in the production process cannot be improved without accurate measurement.

In 1990s, spacecraft sent space telescopes 60 km above the surface of the Earth, freeing the atmosphere from interference with astronomical observations, which was regarded as a new advance in world science and technology at that time. However, the space telescope malfunctioned shortly after liftoff. After several attempts, people finally found out that there was an error of 1 mm in the reflection type zero-value calibrator, which affected the performance of the space telescope and caused unnecessary losses. This is a typical case of product quality degradation due to inaccurate measurement of product characteristic parameters. Research at Rolls-Royce of UK suggests that the process of developing a new type of engine involves a series of metrological measurements. When the uncertainty of dimension measuring instrument is $0.75\,\mu m$, 200 experiments and 20 million dollars are needed. But when the measurement accuracy is improved and the uncertainty is reduced to $0.5\,\mu m$, only 28 experiments and 2.8 million dollars are needed. Hence, accurate measurement brought by metrology can not only save cost and time but also improve product quality.

A large commercial aircraft is composed of millions of components, involving dozens of complex systems such as flight control, hydraulic, electrical, avionics, power fuel and environmental control. Every component requires extremely rigorous precision measurements. Metrological measurement work runs through the whole life cycle of the commercial aircraft from pre-research to design, manufacture, test flight, customer service, repair and maintenance, thus playing an important role in supporting and ensuring. For example, the vibration value of acceleration in aero-engine blade measurement in China's aviation is traceable to a resonant high-acceleration vibration standard equipment developed by National Institute of Metrology, China (NIM) (see Fig. 2.3). It has realized for the first time in the world accurate measurement of the maximum $10,000\,\mathrm{m/s^2}$ vibration acceleration, providing technical support for high-quality production of aero-engine blades in China.

Canada has conducted statistics on the investment efficiency of metrological standard equipment for trade certification, arguing that the ratio of investment to payoff is 1:11, that is, for every Canadian dollar invested in metrological verification, it can avoid the loss of \$11 caused by inaccurate measurements. The statistical research abroad shows that the measurement cost accounts for about 20% of the production cost of communication optical fiber, while in large-scale integrated circuit production, the cost accounts

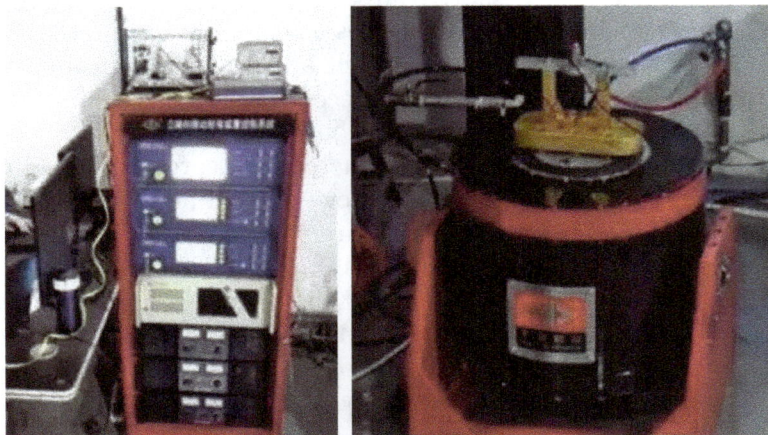

Fig. 2.3.   Resonant high-acceleration vibration standard equipment developed by National Institute of Metrology, China.

for 30%. According to the statistics of America's National Institute of Standards and Technology (NIST), in semiconductor production, the investment-to-payoff ratio of improving the measurement of thermophysical properties is 1:5, while that of improving the measurement of silicon heat resistance reaches 1:37.

From the orderly and sustainable development of a category of industries, from basic research in the laboratory to industrialization in the manufacturing plant, a product has to go through the route of basic scientific research stage, pilot test after maturity and promotion to businesses for mass production. As shown in Fig. 2.4, each constituent element of NQI appears successively in different node places of the whole route. The initial stage of R&D belongs to the technical trial stage where, in order to discover the feasibility of the technology, the result of the relativity measurement is required so that the characterization technique can meet the requirement. With the further development of scientific research, an experimental scheme shall be put forward to evaluate and verify the feasibility of the scheme through the measurement results, thus greatly raising the requirement for the accuracy of measurement results. In order to reduce the number of trial-and-error experiments and improve the efficiency of scientific research and the quality of programs, it is particularly important at this point to improve the accuracy of

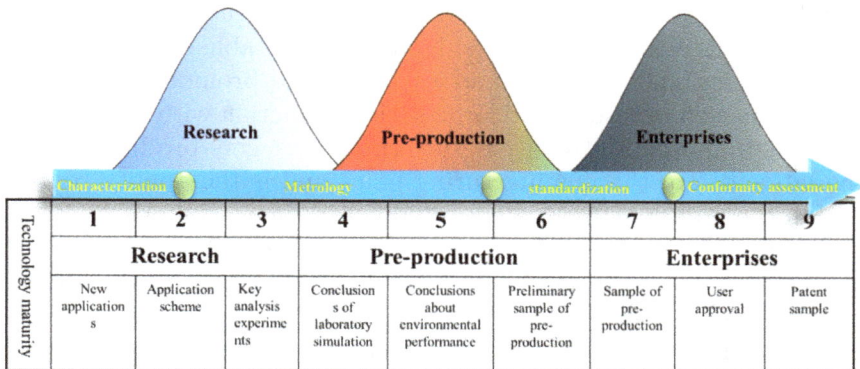

	1	2	3	4	5	6	7	8	9
	Research			Pre-production			Enterprises		
	New applications	Application scheme	Key analysis experiments	Conclusions of laboratory simulation	Conclusions about environmental performance	Preliminary sample of pre-production	Sample of pre-production	User approval	Patent sample

Fig. 2.4. Product R&D, production roadmap and intervention nodes of NQI constituent elements.

measurement results and reduce the measurement uncertainty. As noted earlier in the Rolls-Royce study, metrology needs to intervene in this node and play its supporting role of accurate measurement. The metrological intervention process will continue until the customer's acceptance of product quality in the enterprise's mass production process. When the product has passed the pilot test and is about to enter the large-scale industrial production process, standardization begins, that is, standardizing the measurement results of R&D and pilot test supported by metrology to form standards as a basis for product inspection and quality control, to ensure the consistency of production batches and the control of production quality with a view to guiding large-scale industrial production. Standardization begins with the intervention at the end of the pilot test and continues till customer recognition of product quality. Conformity assessment intervenes in the formation of products during large-scale industrial production, carries out product inspection, testing, certification and accreditation according to pre-formed standards, performs third-party evaluation for the credibility of product quality, and transfers to the market trust in product quality.

Thus, metrology is the basis of all key elements of NQI. In the later stage of basic research, metrology is involved by effective measurement technology, to ensure the reliability and repeatability of measurement results, thereby improving the efficiency of results' transformation. After the successful transformation of research results, on the one hand, metrology controls the quality of the products through testing; on the other hand, it will establish effective measurement method through standardization and promote the orderly transformation of results and large-scale production with the integration of standards. Metrology is the crutch that gets products out of the lab and into the manufacturing plant and is the quality control tool in the orderly, large-scale production in the workshops.

In its accreditation criteria document (CNAS-CL01-G002: 2018) of National Accreditation Committee for China's Approval Assessment (CNAS) "Metrological Traceability Requirements for Measurement Results," CNAS makes this opening statement: "Metrological traceability is a prerequisite for international mutual recognition of measurement results. CNAS regards metrological traceability as the basis for the validity of measurement results, and ensures the metrological traceability of approved measurement activities meet the requirements of international standards." Section 4.1.4 of *General Metrological Terms and Definitions* in State Technical Specifications for Metrology of People's Republic of China (JJF1001-2011) defines metrological traceability as "through the uninterrupted calibration chain prescribed by the document, the results of the measurement are linked to the reference object, and each calibrating in the calibrations chain will introduce measuring uncertainty."

2.4 The Demand for Nanotechnology for Metrological Technology

As is well known, China's basic research papers on nanotechnology are at the forefront of the world in both quantity and quality. The number of related patents is also among the highest in the world, but the industry conversion rate of research results is very low. Apart from the limitations of engineering conditions, a more important bottleneck is that the results of laboratory measurements cannot be reproduced in the process of industrialization.

CNAS explicitly puts forward the requirements of metrological traceability in its accreditation guidelines document CNAS-CL01, *Accreditation criteria for the competence of testing and calibration laboratories* (ISO/IEC17025: 2017). Based on the need to address the accuracy and consistency of laboratory measurement results, the CNAS *Nanotechnology Special Committee was* established in May, 2004, electing Academician Xie Sishen as Chairman of this committee. Accreditation of laboratories is a formal recognition of the ability of calibration and testing laboratories to perform specific types of calibration and testing (this definition is cited from *General Metrological Terms and Definitions*, Section 9.47, JJF1001-2011). The accreditation of nanomeasurement laboratories raises the issue of traceability of equipment values. Metrological technical requirements raised from basic research and industry demand orientation also reflect the technical foundation status of metrology for laboratory accreditation.

Although the establishment of the *Nanotechnology Special Committee* of CNAS has solved part of the problems in laboratory quality management and equipment measurement accuracy, the results still failed to meet the requirements of consistency when different nanomaterials are measured. To solve this problem, in April 2005, Academician Bai Chunli led the establishment of the National Nanotechnology Standardization Technical Committee (SAC/TC279) and served as the first chairman, with the aim of developing high-quality, nanotechnology standards. What are high-quality standards? High-quality standards are embodied in technical standards, which aim to ensure the consistency of measurement results.

In the standardization of nanotechnology, a number of common problems have been discovered in the formulation of equationting technical standards in China. First, the operability of the standard and the consistency of the measurement results need to be improved. The main reason is the lack of laboratory comparison results in the process of establishing technical standards to support and validate standard technical terms, thus ensuring the universality and operability of technical standards. Comparison is a metrology term. Section 4.9 in JJF1001-2011 defines comparison as "the process of comparing the values reproduced by the same measuring instrument for the same accuracy level or for the establishment of uncertainty ranges under specified conditions." Second, high quality proposals from China's nanotechnology standards are required in ISO. International standard proposals like ISO/IEC need to involve more participating countries based on the comparison results of many international laboratories in the international platform for technical support. Our experience with international standard proposals comes from participation in the comparison of Versailles Project on Advanced Materials and Standard (VAMAS). We made the proposal to the technical committee of ISO based on the comparison results, so as to set up the project efficiently and promote the implementation of the standards in various stages.

Metrological institute is the main body leading the organization of international and domestic comparisons, which is able to provide technical solutions, samples and professional competence in data processing and evaluation of comparison results required by *Comparison*. Not only can it provide technical support for parameter confirmation and data consistency in technical standard clauses, but it can also provide the technical basis for international standard proposals by making standard measurement results internationally equivalent through the authoritative comparison platform so that more countries will be invited to participate in the international comparison. The international comparison is usually carried out on the authoritative platform of National Metrological Institutes (NMIs) with its rich experience in this professional field. This fully reflects the technical basis of metrology for standards and the technical support for high-quality standard compilation and fully reflects the dominant position of metrology in the NQI.

In addition, there are still problems of low parameter matching in the synergy of quality infrastructure in China. Standards are the basis for inspection, testing and certification, therefore, the parameters in standards shall be set in accordance with the parameters recognized by the certification requirements. Metrology needs to carry out traceability, comparison of measurement methods based on the requirements of the parameters in the standard, verify the accuracy of measurement by such technical support as traceability, uncertainty evaluation and comparison, meet the need for consistency of measurement results, and finally achieve parameter consistency in measurement and standards, as well as inspection, certification and accreditation, so that the various components of NQI can work harmoniously and efficiently.

SAC/TC279 fully recognizes the decisive role of metrology in the accuracy and consistency of measurement results and support for high-quality standards. Hence, it has chosen to set up the secretariat of the working group on nanotechnology measurement (SAC/TC279/WG5) in NIM, China, contributing to the preparation and release of high-quality standards of nanotechnology and promoting the process of nanometer standardization in China.

Basic research in the field of nanotechnology is almost synchronized with industrial development, which requires that standards can be synchronously supported in the process of transformation of nanoscience into industry. It means that the synchronous standard needs the support of metrological technology to make the measurement results of technical measurement standards equivalent internationally and domestically, and the measurement technology standards universal and operatable.

2.5   Technical Development of NQI for the Industry of Graphene and Related 2D Materials

The Chinese government attaches great importance to the development and application of graphene and related 2D materials (GR2M), laying out plans for the development of GR2M in major national planning documents such as "Made in China 2025," "Outline of the Thirteenth Five-year Plan" for National Economy and Social Development of the People's Republic of China and "Outline of National Innovation-driven Development Strategy."

At present, many leading enterprises have emerged, such as Changzhou No.6 Element Material Technology Co. Ltd., Ningbo Moxi Technology Co. Ltd., Chongqing Moxi Technology Co. Ltd. and Changzhou 2D Carbon Technology Co. Ltd.

With the common drive of all parties, the industrialization of graphene in China has entered the fast track of the synchronization of basic R&D and industrial applications. It can be seen from Fig. 2.4 that now is the best time for the GR2M industry to intervene. Under the guidance of "A high-quality power," metrology begins to intervene in the R&D phase of the GR2M industry, carrying out research on accurate measurement technology, supporting the development and publication of national standards, group standards and international standards, developing performance and inspection method standards for graphene and its products.

In 2018, *Advanced Materials*[10] published an article named "The Worldwide Graphene Flake Productions," which measured and analyzed the parameters of graphene from more than 60 enterprises and concluded that most of the products were not graphene. This article made a good point on how to define graphene and how to call graphene-based products. *Angewandte Chemie International Edition*[11] made a three-dimensional classification map of GR2Ms derived from the single-layer graphene (Fig. 2.5), which were respectively the functional groups in the x axial direction (e.g. C/O), dimension of lateral size in y axial direction and the number of vertical layers (or thickness) in z axial direction. According to the current ISO/TS80004 13:2017 standard for graphene terminology, only a single layer of graphene can be called graphene, different products will be derived differently based on a single layer of graphene because

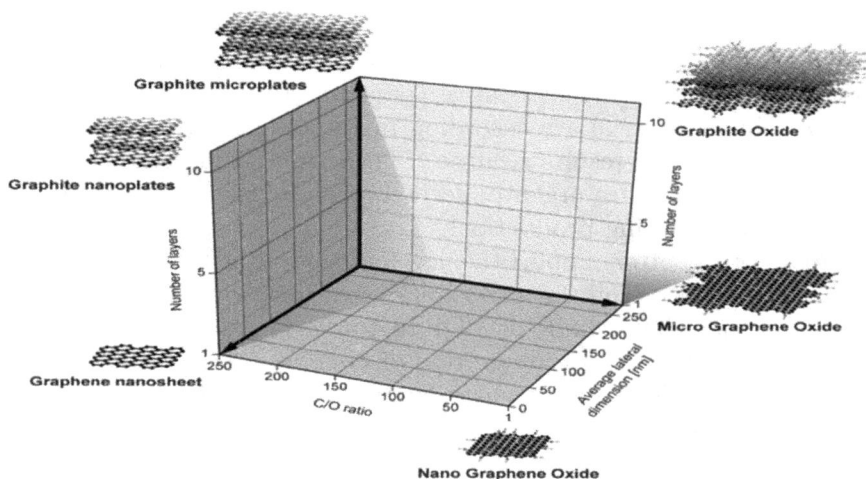

Fig. 2.5. Three-dimensional classification of graphene and related 2D materials.[11]

of their lateral and vertical dimensions and functional groups, suitable for heat conduction, electric conduction, lubrication and other application fields. A generic name for these products was discussed in depth at the annual conference in Hangzhou in November 2019 (ISO/TC 229) and tentatively named them Graphene and Related Two-Dimensional Material (GR2M) to meet the market demand for trade, which was not definite in the standard ISO/TS 80004-13:2017. Hence, most of the graphene-related products claimed by more than 60 graphene enterprises in the literature[10] are not graphene. This is in contradiction with our enterprise's claimed graphene products because, in our industry, the various horizontal vertical sizes various graphene-based products based on various size, thickness and functional groups within 10 layers are called graphene materials, but this name is not accepted internationally because it was not defined in the ISO standard. It can be seen from different examples of the paper and industry how important term standards are. As a result, how to give a generic definition of graphene-based products that are accepted by both the scientific and industrial communities internationally is crucial to market confidence and industrial development. Only a clear, simple and complete definition is conducive to the scientific popularization of this type of material and can make the market carry on the product trade better and more clearly. Other issues such as the current company's claims to manufacture or use the graphene material are according to what claims? How to detect them? Are the test results accurate? These are common issues of the quality bases that are urgently needed to be solved in the R&D and design of engineering productions.

In order to better support the development of the GR2M industry, researchers have taken the conformity assessment requirement as the guide in the whole chain of NQI to top-level design on the NQI technology research and integration of GR2Ms based on the above-mentioned problems in the scientific and industrial fields. Finally, a road map of the standards and metrologies of GR2Ms is built up and shows the urgent problems to be solved in the GR2M industry.

As is shown in Fig. 2.6, the market demand for GR2Ms in China is analyzed that domestic GR2Ms have two ways: one is GR2M (powder) prepared by chemical oxidation and reduction oxidation and physical ball milling and the other is GR2M (thin film) prepared by Chemical Vapor Deposition (CVD). The requirements are relatively high for production technology and process of graphene film materials. The production of a product in the market is only a matter of quality. At present, graphene powder products are the

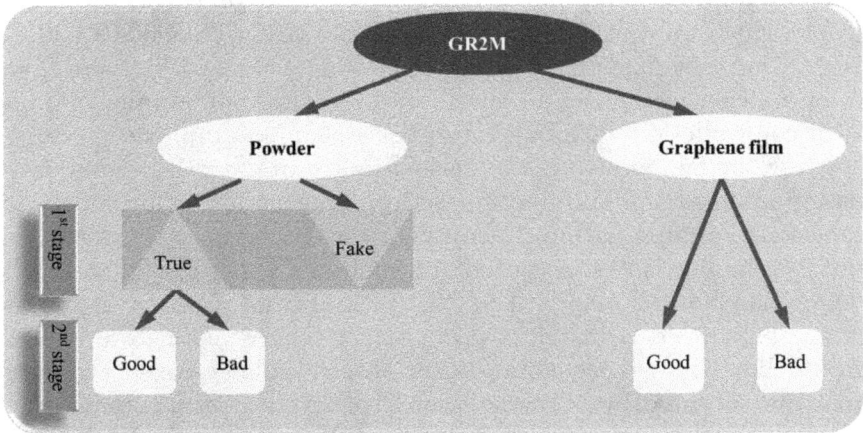

Fig. 2.6.   Market demand analysis of GR2M in China.

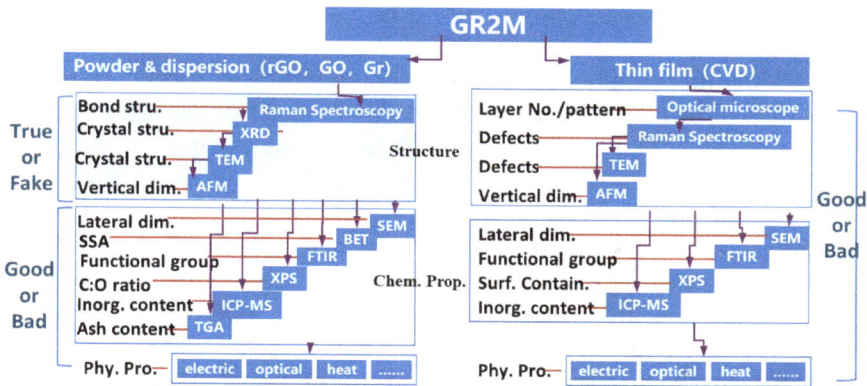

Fig. 2.7. Road map of the metrological technologies of GR2M.

main body in China GR2M's market. In the market trade, the different levels of production technology have caused high or low quality of GR2M products. Moreover, it is difficult to distinguish true or fake GR2M based on GR2M term. Therefore, it is necessary to establish effective measurement technology to evaluate the final product.

Top-level design for the metrological research is carried out according to the common problems of quality basis of GR2M, as shown in Fig. 2.7. It is clear from Fig. 2.7 that studies of metrological technology of GR2M include the calibration or traceability of equipment and establishment of valid measurement methods for different measurement parameters. Whether it is the powder of GR2M or the film of GR2M, the first thing for the metrology is to do the calibration or traceability of the related equipment used like Raman spectroscopy, Transmission Electron Microscope (TEM), Scanning Electron Microscope (SEM), Atomic Force Microscope (AFM), X-ray Diffractometer (XRD), Optical Microscope (OM), X-ray Photoelectron Spectroscopy (XPS), Inductively Coupled Plasma Mass Spectrometry (ICP MS), Fourier Transform Infrared Spectrometer (FTIR), Brunauer–Emmett–Teller (BET), Themo Gravimetric Analyzer (TGA), thermo-electro-optic and other performance measurement equipment. The aim of the above traceability research is to ensure the accuracy and reliability of the measurement equipment. After the calibration of the equipment, the next thing is to study the valid measurement method of each parameter.

What is a valid measurement? The valid measurement includes accurate equipment and reliable measurement method. Meanwhile, the reliable measurement method is reliable on measurement conditions and data analysis. The following is a brief explanation of the measurement of GR2M layers as is known to all; a single theoretical layer thickness of graphene is 0.334 nm and the thickness of three layers is only about 1 nm. Accurate measurements in such a small range of dimensions raises higher requirements for both measuring equipment and measuring methods. AFM is currently recognized as one of the absolute measurement methods. On the one hand, the lower limit of measurement for AFM equipment shall be below 1 nm and measured results shall be traceable to the SI basic unit. NIM supplied the certified reference materials of single atomic step height of strontium titanate, which can be used to calibrate AFM around 1 nm scale. On the other hand the measurement method shall be reliable within such a small range of 1 nm. Figure 2.8 shows thickness measurement of graphene oxide by AFM. It is clear from the right figure that the error of the step baseline in the red circle due to contaminants and instrument noise has reached nearly 1 nm. Ensuring the measurement reliability of around 1 nm thickness is the biggest challenge facing the layer measurement of GR2M. NIM provides a method of data analysis and leads an international comparison to ensure the measurement result is consistent and reliable. Thus, solving the problem of accurate and reliable measurement results of GR2M parameters from

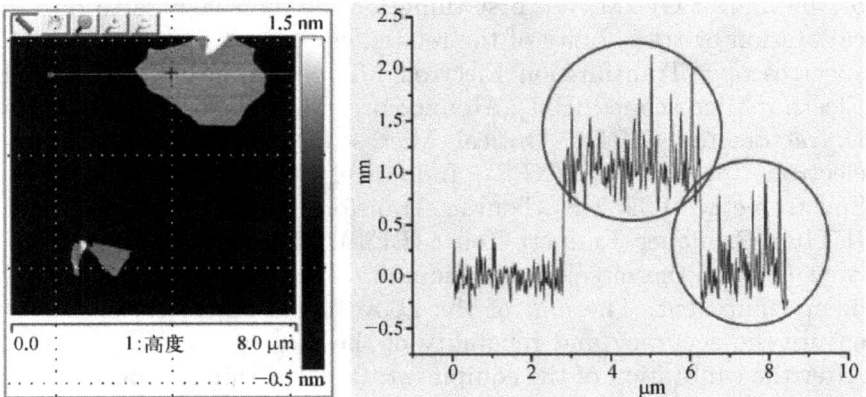

Fig. 2.8.   Image and figure of graphene oxide thickness measurement using AFM.

structure, composition and properties is the most pressing measurement need of the industry. Please refer to Chapters 3–7 for technical details.

Similarly, these accurate and reliable measurement results of GR2M parameters are the base for product authenticity. Moreover, the assessment process combined with these reliable results of GR2M parameters is the soul of product quality assessment due to the complexity of GR2Ms. For example, the structural parameters are confirmed for the assessment of true or fake GR2M, which are characterized using Raman spectroscopy, XRD, AFM and TEM; the assessment process with the measurement order is confirmed: First, use Raman spectroscopy to characterize the chemical valence bond structure to prove that it has basically a sp^2 hybridized C=C honeycomb structure. If yes, then use XRD to characterize the crystal structure to prove that it has a graphite-like or graphite oxide crystal structure. If yes, TEM and AFM are used to further confirm the morphology, number of layers and thickness of the sample from a microscopic view. When we complete four-step characterizations, all of the obtained results are combined to make a conclusion of what kind of GR2Ms they are. As mentioned above, metrological technological achievements and method standards both support the judging of product authenticity and ensure confidence in market trade.

However, end users have different requirements for the quality level of the products. The product manufacturer needs to meet the consistency of different production batches, while the product user needs to satisfy the functional needs, such as electrical conductivity, thermal conductivity and anti-corrosion. Therefore, both the manufacturer and user shall pick out specific chemical and physical parameters according to the domain product application and direction for assessments of production quality and obtain reliable measurement results and valid method standards of these selected parameters. Only in this way can we make a conclusion about the consistency of production batch and the quality of application effect of the product.

From GR2M metrology, materials metrology is not only the work on the traceability or calibration of measurement equipment but also the key parameters' selection and their valid measurement, moreover, the critical values for material quality assessment, which helps follow up the development of systematic standards and helps make

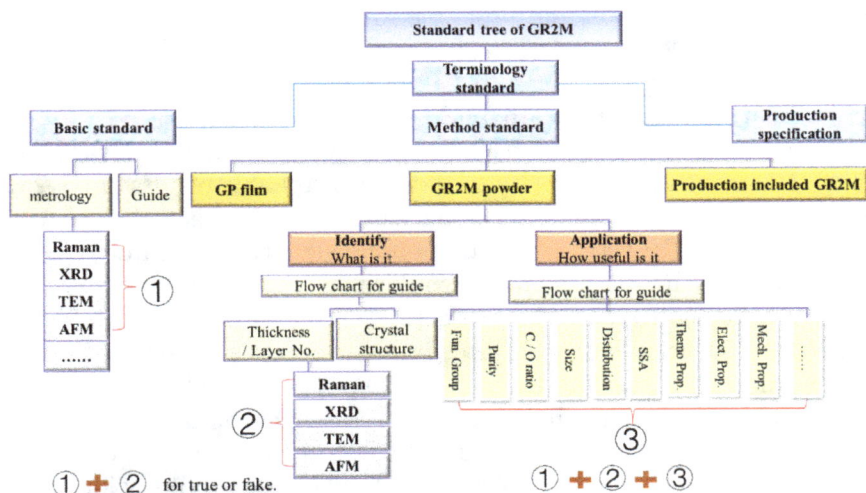

Fig. 2.9.   The top layer design of standard system for GR2Ms.

the adoption of standards systemic, thus playing a quality-based role in product quality assurance and improvement and orderly trade.

Thus, the framework of GR2M standardization was carried out based on the road map of GR2M metrology, as shown in Fig. 2.9. In Fig. 2.9, term standard is the first level while guide and metrology, method standard and product specification are the second level in the framework of GR2M standardization. Meanwhile, both ① and ② are related to equipment, however, ① is the standard system for measuring equipment calibration or traceability, and ② is the standard system of measurement methods using the measuring equipment. At present, NIM has completed the equipment calibration in ① and published calibration specifications. Subsequently, NIM metrologists published method standards of parameters of chemical structure, crystal structure and layer numbers of GR2M using Raman spectroscopy, XRD, AFM and TEM with attention of metrologists to detail and committed to the global comparability mindset.[4] These standards have achieved mutual recognition of international equivalence through international comparisons and five standards are published under Chinese Society for Testing and Materials (CSTM), as shown in Table 2.1. Supported by the test results using five CSTM standards, the first graphene product certification certificate (Fig. 2.10) was issued at the "Graphene 2018" Global Graphene

Table 2.1. Published five group standards of GR2M powder under CSTM.

No.	Standard no.	Standard name
1	T/CSTM00166.1-2019	Characterization of GR2M, Part I: Raman spectroscopy
2	T/CSTM00166.2-2019	Characterization of GR2M powder, Part II: X-ray diffraction
3	T/CSTM00166.3-2019	Characterization of GR2M, Part III: Transmission electron microscope
4	T/CSTM00003-2019	Thickness measurement of two-dimensional material by atomic force microscopy
5	T/CSTM00168-2019	The determination specification of GR2M powder

Spring Conference, held in Dresden, Germany, on June 28, 2018. It's a typical example that the integration of NQI components and the implementation of full chain applications in the graphene industry will promote the healthy development of the graphene industry through the inspection and conformity assessment process of the factory.

The world's first "GR2M product" certification proved again that the reliable measurement method of the key multi-parameters of the GR2M powder and slurry has provided the market with the "eye" to distinguish product authenticity. It further explains that metrology is the benchmark of product quality.

"Measurement value" is the first element in all NQI technological constituents. The "measurement value equivalence" is a final objective and is the key to the integration of NQI chain. "Measurement uncertainty narrowing" is a means of quality improvement. At the different stages of industry, the emphases of NQI technical constituents are different, which need to launch force sequentially under overall management. The whole chain implementation of NQI technical constituents provides a channel for metrology to serve the society, making industrial quality assurance and promotion possible. It provides a guarantee for the regulation of market behavior, promoting market circulation and improving market efficiency using metrological mechanism, significantly lowering market costs for market transactions through confidence mechanisms.

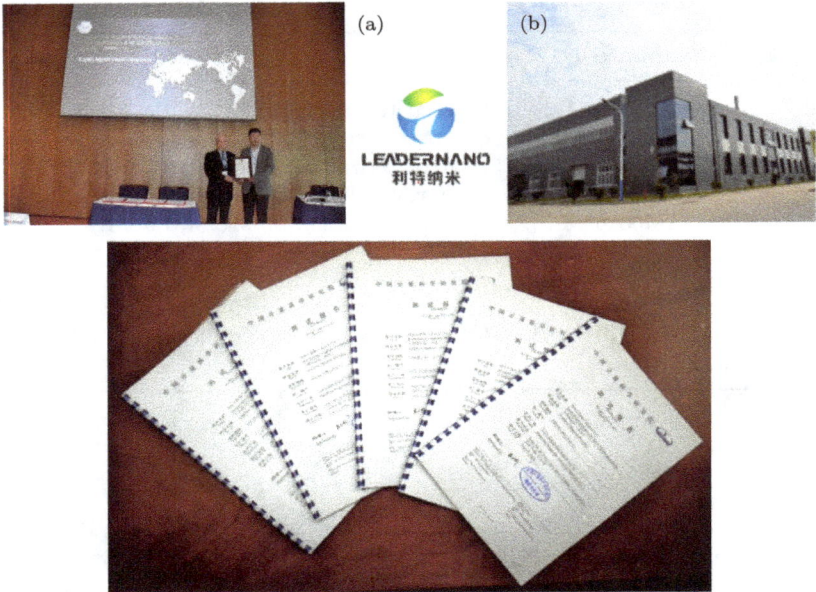

Fig. 2.10.   Liu Zhongfan, Academician of Chinese Academy of Sciences received on behalf of Shandong Li-Te Nanotechnology Co., Ltd. the world's first "GR2M" product certification issued by International Graphene Product Certification Center (a) based on the testing reports from NIM (b).

From Chapter 3 onwards, we elaborate on the key elements of the valid measurements of key parameters of GR2Ms: equipment calibration and traceability, development of related reference materials, international comparison in the process of measurement method standardizations and development of international standards.

References

[1]   JJF 1001–2011. *General Terms in Metrology and Their definitions* [S].
[2]   GB/T 20000.1–2002. *Guide for standardization-Part 1: Standardization and related activities-General vocabulary* [S].
[3]   GB/T 27000–2006. *Conformity assessment-Vocabulary and general principles* [S].
[4]   Sene M, Gilmore I, Janssen J T. Metrology is key to reproducing results. *Nature*, 2017, 547(7664): 397–399.

[5] Quinn T. *From Artefacts to Atoms — The BIPM and the Search for Ultimate Measurement Standards.* Zhang, Yukuan (trans). Beijing: China Quality and Standards Publishing & Media Co., Ltd., 2015.

[6] GB/T30544.13–2018 (Equivalent to ISO/TS 80004–13:2017). *Nanotechnologies-Vocabulary-Part 13: Graphene and related two-dimensional (2D) Materials.*

[7] JJF1059.1–2012. *Evaluation and Expression of Uncertainty in Measurement* [S].

[8] Ni Y. *Evaluation of Uncertainty in Practical Measurement* (3rd edn.). Beijing: China Metrology Publishing House, 2009.

[9] JJF 1117–2010. *Measurement Comparison* [S].

[10] Kauling A P, Seefeldt A T, Pisoni D P, *et al.* The worldwide Graphene flake production. *Advanced Materials*, 2018, 30(44): 1803784.

[11] Wick P, Louw-Gaume A E, Kucki M, *et al.* Classification framework for Graphene-based materials. *Angewandte Chemie International Edition*, 2014, 53(30): 7714–7718.

https://doi.org/10.1142/9789811295713_0003

Chapter 3

Metrological Technology of Raman Spectroscopy for Graphene-Related Materials

Yaxuan Yao

3.1 Overview

Raman spectroscopy is one of the most commonly used, fast, non-destructive and high-resolution techniques for characterizing graphene-related materials. Excited by an incident laser, the electrons in the valence band of graphene-related materials transit to the conduction band, and the electron–phonon interaction scatters, thus producing different Raman bands. Characteristic Raman bands of graphene-related materials include D band, G band and 2D band (G′ band). Figure 3.1 shows Raman spectra of single-layer to few-layer graphene-related materials prepared by different methods.[1–3] D band usually appears in the vicinity of 1,350 cm^{-1}; it is caused by radial breathing pattern of symmetrical stretching vibration of sp^2 carbon atoms in the aromatic ring and usually requires the presence of a defect to activate. G band mainly appears in the vicinity of 1,580 cm^{-1}; it is caused by the stretching vibration between sp^2 carbon atoms, corresponding to the vibration of optical phonon E_{2g} at the center of Brillouin area. 2D band is a frequency-doubling band of D band, usually appearing in the vicinity of 2,680 cm^{-1}. It is caused by double resonance transition of two phonons with opposite momentum in a carbon atom.

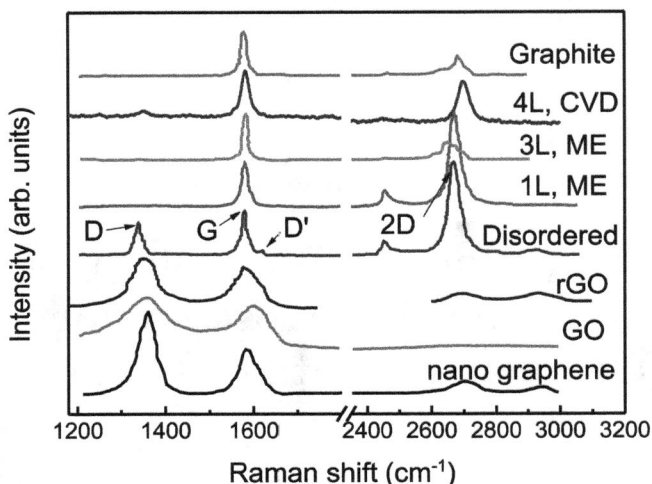

Fig. 3.1.   Raman spectra of single-layer to few-layer graphene-related materials prepared by different methods.[1-3]
Note: L: Layer; CVD: Chemical vapor deposition; ME: Mechanical exfoliation; rGO: Reduced graphene oxide; GO: Graphene oxide.

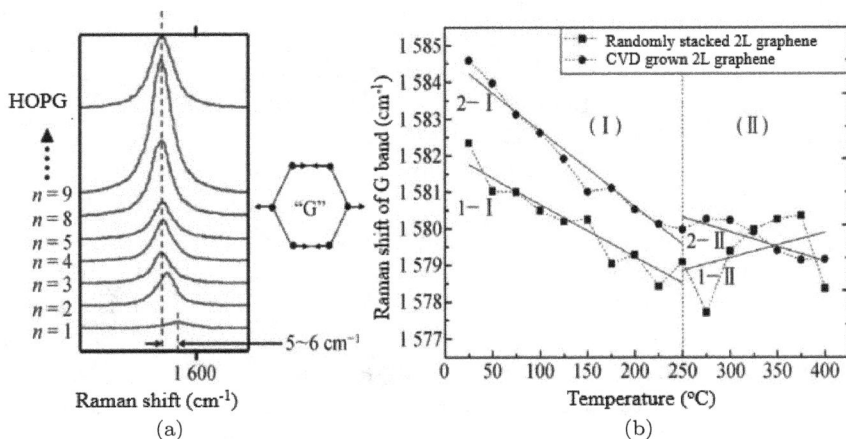

Fig. 3.2.   Changes of G band position of graphene[4,5]: (a) changes of G band position with the number of layers; (b) changes of G band position with the temperature (°C).

The positions of characteristic Raman bands are often used to qualitatively characterize the number of layers, doping and stacking of graphene-related materials.[4-6] Figure 3.2(a) shows the changes of

G band position with the number of layers. For graphene prepared by mechanical exfoliation, G band of single-layer graphene appears in the vicinity of $1,587 \, cm^{-1}$. With the increasing of layer numbers, the intensity of G band increases, while the position of G band shifts to lower wavenumber. The difference in G band position between single-layer graphene and highly oriented pyrolytic graphite is (5–6) cm^{-1}. Figure 3.2(b) shows temperature dependence of G band position of randomly stacked double-layer graphene and CVD-grown double-layer graphene. Generally, G band position of graphene moves linearly to lower wavenumber with the increase in temperature, which is probably due to the extension of C–C elongation caused by thermal expansion or anharmonic phonon coupling, thus reducing force constant, softening in-plane optical phonons and lowering vibration frequency. For double-layer graphene grown by CVD method, as the temperature rises from room temperature to 250°C, G band position shifts from $1,585 \, cm^{-1}$ to $1,580 \, cm^{-1}$, with a difference of (5–6) cm^{-1}. Therefore, accurate measurement of Raman shift is helpful to study and analyze the structure of graphene-related materials.

The relative intensity of characteristic Raman bands is often used to study the number of layers and defects of graphene-related materials. For example, Full Width at Half Maximum (FWHM) of 2D band, intensity ratio and peak area ratio of G band to 2D band are often used to determine the layer number of graphene-related materials in AB stacks.[1,7] Figure 3.3(a) shows Raman spectra of 1–4 layers of graphene-related materials prepared by mechanical exfoliation. When FWHM of 2D band is about $30 \, cm^{-1}$ and the intensity ratio of G band to 2D band is $I_G/I_{2D} < 0.7$, the sample can be determined as single-layer graphene. When FWHM of 2D band is about $50 \, cm^{-1}$ and $0.7 < I_G/< I_{2D} < 1$, the sample can be determined as double-layer graphene. For another example, the intensity ratio of D band to G band is an effective method to study defect type and defect density of graphene-related materials. D band stands for vibrational breathing modes of sp^2-hybrid carbon ring in graphene, which shows the defect and disorder degree of carbon lattice. In graphite and high-quality graphene, D band is generally relative weak. D band intensity is proportional to the degree of defects in the sample. Figure 3.3(b) shows Raman spectra of single-layer graphene with different defect concentrations. As the concentration of graphene defects increases, the intensity of D band increases, intensity ratio of D band to G band

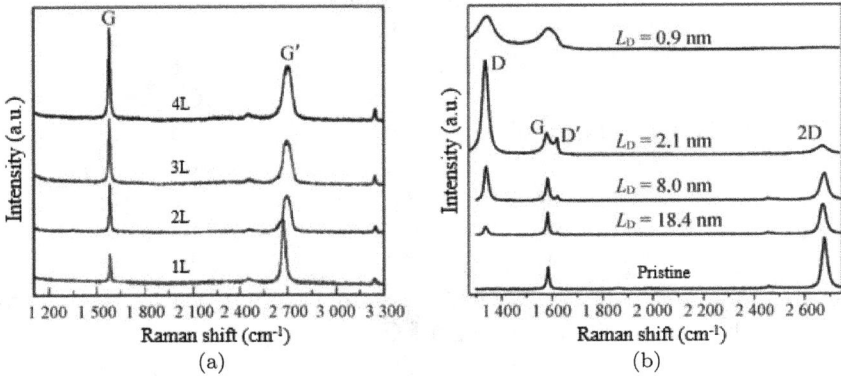

Fig. 3.3.   Relative intensity changes of characteristic Raman bands of graphene-related materials[6,7]: (a) Raman spectra of 1–4L graphene-related materials prepared by mechanical exfoliation; (b) Raman spectra of single-layer graphene with different defect concentrations.

I_D/I_G keeps increasing, and a relatively weak D′ band appears in the vicinity of $1,620\,\text{cm}^{-1}$. D band and D′ band occur respectively in the process of intervalley and in-valley scattering, and intensity ratio I_D/I_G is closely related to the defect type on graphene surface. When the defect concentration is low, intensity of both D band and D′ band increases with the defect density, proportional to defect density. When the defect concentration increases to a certain extent, the intensity of D band reaches the maximum and then begins to decrease, while the intensity of D′ band remains the same. Therefore, accurate measurement of Raman relative intensity is helpful in studying and analyzing the defect of graphene-related materials.

When the intensity ratio of characteristic Raman bands of graphene-related materials is used, the relative intensity of characteristic Raman bands shall be calibrated to ensure the accuracy and comparability of measurement results. Figure 3.4 and Table 3.1 present the differences in intensity ratio of characteristic Raman bands of graphene before and after calibration, respectively. When the positions of two characteristic Raman bands are close, such as D band and G band, the intensity ratio of these two bands has little or no change before and after calibration. When the positions of two characteristic Raman bands are far apart, such as G band and 2D band, the intensity ratio of these two bands changes significantly

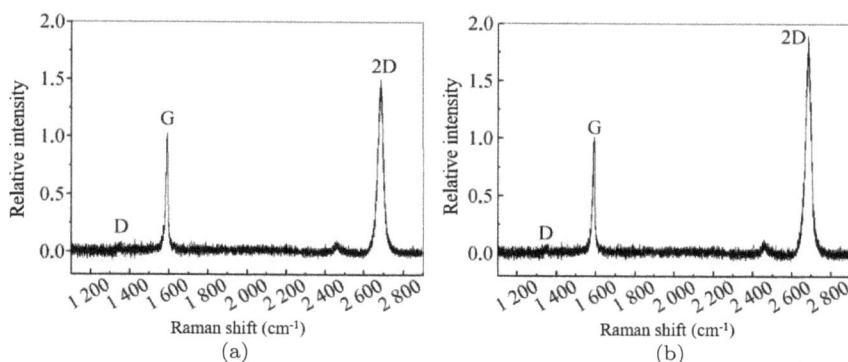

Fig. 3.4.   Differences in relative intensity and intensity ratio I_D/I_G of character-istic Raman bands of graphene before and after calibration (normalized with the intensity of G band): (a) before calibration; (b) after calibration.

Table 3.1.   Differences in intensity ratio of characteristic Raman bands of graphene before and after calibration.

	Before calibration	After calibration
I_D/I_G	0.15	0.14
I_{2D}/I_G	1.55	1.94

before and after calibration. Therefore, it is necessary to calibrate the measured Raman spectra, especially the Raman spectra in a broad range. The calibrated spectra can better reflect actual information of tested samples.

In summary, Raman spectroscopy is an effective technique for characterizing the structure of graphene-related materials. The shape, intensity and position of characteristic Raman bands can be used to qualitatively characterize layer number, defects, doping and stacking of graphene-related materials. Accuracy and reliability of Raman shift and relative intensity are essential for quality control in the production of graphene-related materials, which can provide technical guidance for the exploration and development of graphene-related materials.

This chapter elaborates on measurement principles of Raman spectroscopy, traceability of Raman shift and relative intensity, development of related certified reference materials, calibration methods for Raman spectroscopy through the use of certified reference materials, and measurement methods of graphene-related materials by calibrated Raman spectroscopy.

3.2   Traceability of Raman Spectroscopy

Traceability is the guarantee of accurate and reliable measurement results. *Requirements on the Metrological Traceability of Measurement Results* (CNAS CL01 G002:2018) dictates that only the measurand value which can be linked to national and international metrological standards by means of uncertainty can be regarded as a credible, consistent and comparable result.

Traceability refers to the process of characteristic measured values being linked to specified reference standards, usually national or international metrological standards (benchmarks), through the use of a continuous chain of comparisons with specified uncertainties. Value transfer is the reverse process of traceability. Through verification or calibration of measurement instruments, characteristic value reproduced by national metrological standard is transferred to measurement instruments via various levels of measurement standards, in order to ensure accuracy and consistency of measurand value. Metrological standard is an important intermediate to realize traceability and value transfer of measurand value. On one hand, the establishment of a metrological standard requires traceability of measurand values to a higher level of metrological standards. On the other hand, the reproduced value is transferred to a working measurement instrument through verification and calibration, which can ensure accuracy and reliability of measurement results. Traceability study of metrological standard and development of reference material are generally conducted by metrological institutes, such as National Institute of Metrology, China. Calibration of measurement instruments shall be carried out by metrological institutions or by laboratories themselves to ensure the accuracy and reliability of measurement results.

For Raman spectroscopy, the process of characteristic values traced to optical standard through reasonable and effective procedures is known as traceability process. The Raman spectroscopy which is traced to SI units could be used as a metrological standard through specific assessment procedures, and standard values of reference materials can be determined by the metrological standard. Reference materials can be used for Raman spectroscopy calibration for daily use in various laboratories, thus establishing an uninterruptible trace chain for Raman spectroscopy to SI units.

3.2.1 Measurement Principle of Raman Spectroscopy

Raman spectrometer generally consists of four parts: excitation light source, sample optical path, spectrometer (*spectrograph*) and detector; its basic structure is shown in Fig. 3.5.[8] The laser reaches the sample surface to excite the sample; the scattered light from the excitation is collected, enters the spectrometer for light dispersion and finally reaches the detector to detect Raman signal. There are different types of Raman spectroscopy depending on different dispersion ways of scattering Raman light: filter Raman spectroscopy, Fourier transform Raman spectroscopy and dispersive Raman spectrometer.

Raman scattering is an inelastic scattering process in which energy is exchanged between photons and molecules, as shown in Fig. 3.6. The electrons of molecules in the ground state are irradiated by an incident light with an energy of $h\nu_0$ and transited to an excited virtual state. The electrons transit from the excited virtual state to the next state and emit light, that is, scattered light.

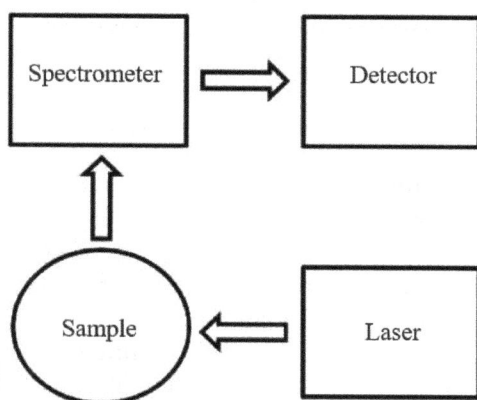

Fig. 3.5. Schematic diagram of basic structure of Raman spectroscopy.

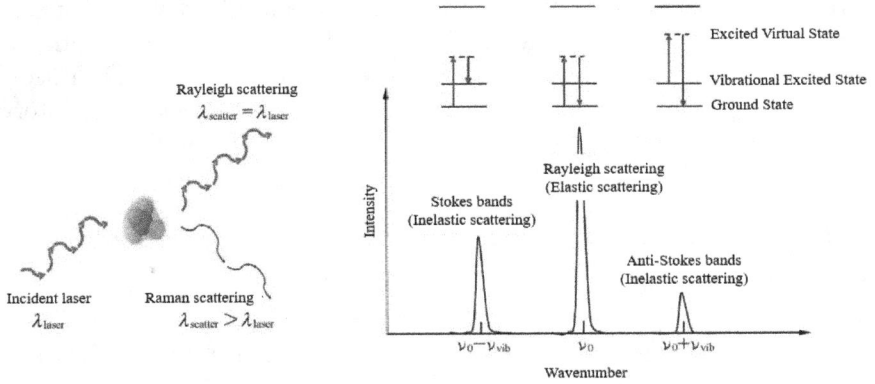

Fig. 3.6.   Diagrammatic sketch of Raman scattering.

The process in which a photon changes only its direction of motion without changing its frequency is called Rayleigh scattering. Due to the exchange of energy between photons and molecules, the process by which the frequency of the scattered light changes is called Raman scattering. The difference $\Delta\nu$ between the frequency of incident light and that of Raman scattered light is called Raman shift:

$$\Delta\nu = \nu_0 - \nu_{vib} = \frac{1}{\lambda_{\text{laser}}} - \frac{1}{\lambda_{\text{scatter}}} \tag{3.1}$$

where λ_{laser} is the wavelength of incident light (nm) and λ_{scatter} is the wavelength of Raman scattered light (nm).

The relative intensity of collected Raman scattered light is expressed as follows:[9]

$$S'_s = P_{\text{D}}\beta'_s s D_s \Omega T A_D Q K t \tag{3.2}$$

where S'_s is the number of collected photoelectrons in relative wavenumber (photonelectron $\cdot$ (cm^{-1})$^{-1}$), P_{D} is the laser density at which the laser reaches the sample (photon $\cdot$ s$^{-1}\cdot$ cm^{-2}), β'_s is the differential cross-section of each molecule surface with respect to relative wavenumber (cm$^2 \cdot$ molecule$^{-1} \cdot$ sr$^{-1} \cdot$ (cm^{-1})$^{-1}$), s is the slit bandwidth (cm^{-1}), Ds is the molecular number density of the sample (molecule $\cdot$ cm^{-3}), Ω is the collection angle (sr^{-1}), T is the transmittance of spectrometer and optical devices (%), A_{D} is the sample area entering the spectrometer (cm^2), Q is the quantum efficiency of detector (photonelectron · photon^{-1}), K is the geometric factor and t is the accumulation time (s).

For the same instrument, the values of $s, \Omega, T, A_\mathrm{D}$ and Q remain constant under identical measurement conditions, defined as

$$R = s\Omega T A_\mathrm{D} Q \qquad (3.3)$$

R also represents the instrumental response function.

In summary, there are two characteristic values of the Raman spectrum, namely Raman shift and relative intensity. Therefore, it is necessary to trace Raman shift and relative intensity respectively, that is, using standard equipment of traceability to trace to the national standard (benchmark). Sections 3.2.2 and 3.2.3 describe the traceability of Raman shift and relative intensity, respectively.

3.2.2 Traceability of Raman shift

Traceability study of Raman shift is based on Equation (3.1). From the equation, Raman shift is related to the wavenumber (reciprocal of wavelength) difference between incident and scattered light; a small change of wavelength of incident light and scattered light may lead to a large deviation of Raman shift. Therefore, the wavelength of incident and scattered light needs to be calibrated and traced.

The calibration and traceability of scattered light wavelength is also called optical path alignment or spectrometer calibration. According to geometrical optics, the more parallel the beam is projected onto the grating, the more evenly the grating can be illuminated. Through the dispersion of grating and the focusing of the reflector, the higher image quality will be received on the CCD detector, resulting in more accurate and reliable measurement results. Thus, a beam of light that does not shift between incident light and emit light is considered as an optical path alignment, that is, the wavelength of incident light equals to detected wavelength. A standard light source with known wavelength can be used for wavelength traceability and calibration of scattered light.

When the measured signal of the standard light source after passing through the optical path of the spectrometer is collected, traceability and calibration of scattered light can be achieved by adjusting the spectrometer until the measured wavelength equals that of the

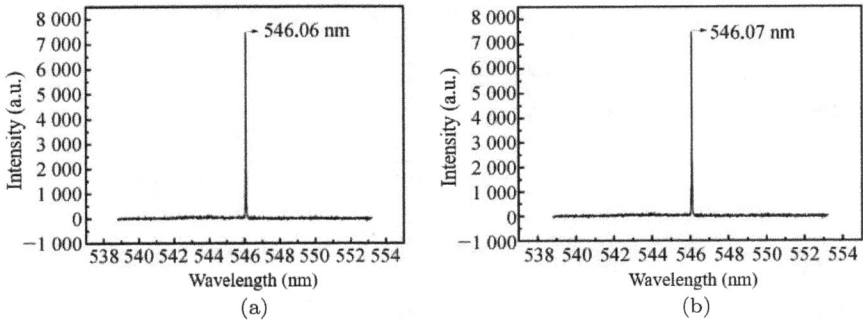

Fig. 3.7.   Traceability process of Raman spectrometer: (a) before tracing; (b) after tracing.

standard spectral line. Figure 3.7 shows the traceability process of the Raman spectrometer. Based on the performance of grating in the spectrometer, the standard spectral line nearest to the blazed wavelength of grating can be selected as the trace source of the spectral line. For example, if the blazed wavelength of grating is in the vicinity of 500 nm, a spectral line at 546.07 nm of a standard mercury argon lamp can be used. When measuring the spectral line of a standard mercury argon lamp at 546.07 nm (standard value), if the measurand value does not conform with the standard value (Fig. 3.7(a)), optical path parameters shall be adjusted until the measured value is consistent with the standard value (Fig. 3.7(b)), thus the traceability process of Raman spectrometer has been completed.

The wavelength calibration and traceability of incident light is performed at $\Delta \nu = 0$ according to the principle of Raman scattering. As Rayleigh scattering and residual lasers cannot be completely filtered out, a portion of the optical signal with the same wavelength as incident light λ_{laser} still enters the spectrometer along with the Raman scattering signal and participates in the follow-up process. From Equation (3.1), Raman shift shall be 0 for emit light with a λ_{laser} wavelength. For the wavelength of the laser (incident light) calibration and traceability, a possible wavelength value of incident light can be input directly into the software, and the spectrum of the samples at location $\Delta \nu = 0$ is measured. Modify the possible input parameter and repeat this step until the characteristic Raman band reaches its maximum at $\Delta \nu = 0$. At this point, the input wavelength value is the current laser excitation wavelength, and laser wavelength calibration has been completed.[8]

3.2.3 Traceability of Relative Intensity of Raman Spectroscopy

Due to the unique combination of filter, grating, light collecting system and detection system, each spectrometer has its own instrumental response function. Raman spectra measured by uncalibrated spectroscopy may be deformed or distorted, thus reducing the accuracy of measurement results. Besides, the measurement of various instruments by various researchers may cause incomparable results. In addition, when parts are replaced or repaired, Raman spectra measured before and after maintenance also lack comparability.

While ensuring the accuracy of Raman shift, traceability of relative intensity is primarily traceability of the number of photoelectrons collected by Raman spectroscopy. In Equation (3.2), P_D and t are related to measurement condition and β_s' is related to the tested sample. And for the same instrument, when the same measurement condition is selected, the values of s, Ω, T, A_D and Q remain constant. Thus, the traceability of relative intensity could be simplified to the traceability of instrumental response function R. Different from Raman shift, a continuous spectrum with certain photoelectron response value at each Raman shift in the full spectrum is needed for relative intensity. Therefore, it is necessary to select a standard light source of a continuous spectrum as the measurement standard.

When the standard light source was used, the spectrum of the standard light source is represented by L_L, the spectrum obtained by Raman spectroscopy is represented by S_L', and Equation (3.2) can be rewritten as

$$S_L' = L_L s \Omega T A_D Q t = L_L R t \tag{3.4}$$

Therefore,

$$R = \frac{1}{t} \times \frac{S_L'}{L_L} \tag{3.5}$$

Since accumulation time t is a constant which does not affect the shape and relative intensity of the curve, instrumental response function R can be obtained from S_L'/L_L.

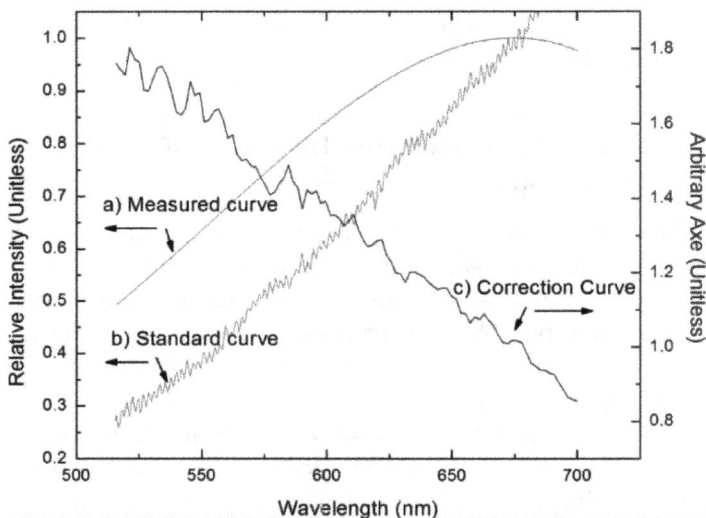

Fig. 3.8. Traceability process of relative intensity for Raman spectroscopy.[11]

Figure 3.8 shows the traceability process of relative intensity for Raman spectroscopy. Through the use of the standard light source, the relative intensity spectrum is collected in the full spectrum range (measured spectrum S'_L, curve a). Dividing the measured spectrum (S'_L, curve a) by the standard spectrum (L_L, curve b), instrumental response function R (curve c) can be obtained. Thus, traceability of relative intensity for Raman Spectroscopy has been completed.[10]

In summary, there are two characteristic parameters in Raman spectrum, namely Raman shift and relative intensity. These two characteristic parameters are both related to optical characteristics; therefore, standard light sources are used in the traceability process. For Raman shift which is related to scattered wavelength, mercury argon lamp, neon lamp and other standard sources with sharp spectral lines shall be used. For relative intensity which correlates with the number of photoelectrons, a standard light source with a continuous spectrum, such as a tungsten halogen lamp, shall be used. The trace road of Raman spectroscopy is shown in Fig. 3.9.

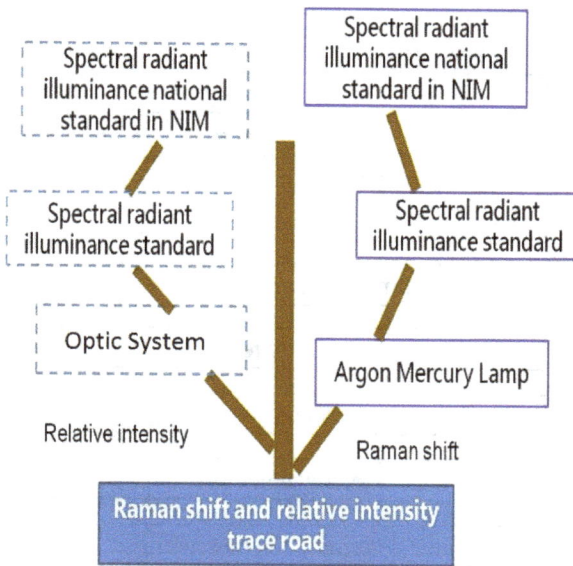

Fig. 3.9. Trace road of Raman shift and relative intensity for Raman spectroscopy.

3.3   Certified Reference Materials of Raman Shift and Relative Intensity for Raman Spectroscopy

From Section 3.2, standard light sources are expensive and complicated to operate, which are only applicable to traceability study in metrological institutions. Certified reference materials (CRM) of Raman shift and relative intensity for Raman spectroscopy are low cost, easy to operate and convenient to carry. These CRMs are necessary for end-users to ensure accuracy and reliability of measurement results of Raman spectroscopy in daily use. Therefore, development of CRMs has also become a focus of the work in metrological study.

CRMs are high-tech measurement references; they are generally developed by national institutes of metrology, metrological laboratories in industry and specialized laboratories in enterprises or universities. The development process of CRMs is shown in Fig. 3.10.[11] General steps are needs assessment, feasibility study, selection of

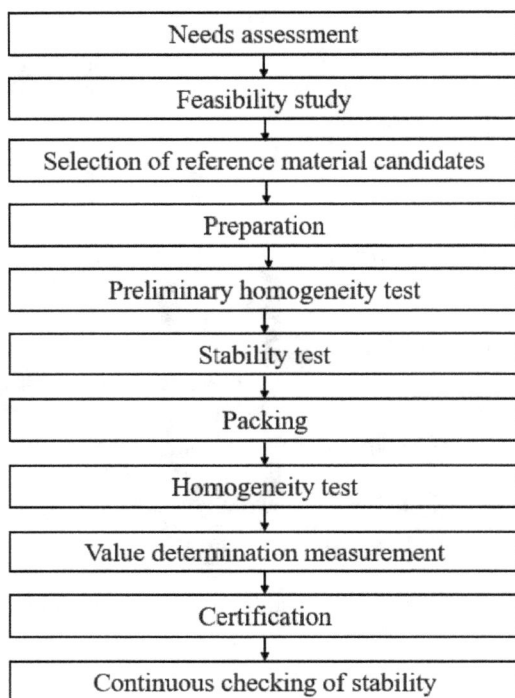

Fig. 3.10.   Development process of certified reference materials.[12]

reference material candidates, preparation, preliminary homogeneity test, stability test, packing, homogeneity test, value determination measurement, certification and continuous checking of stability. With regard to development of Raman shift and relative intensity CRMs for Raman spectroscopy, this section describes homogeneity, stability testing and certification procedures in the process of developing CRMs.

3.3.1   Certified Reference Materials of Raman Shift

There are many options of CRMs for Raman shift calibration. At present, the most commonly used material is a single-crystal silicon wafer with a theoretical Raman shift value of $520.7\,\mathrm{cm^{-1}}$. However, surface damage, orientation and other reasons have caused poor reproducibility of the theoretical Raman shift value of single-crystal silicon wafers. Therefore, it is only suitable for fast inspection of the instrument's daily use. More importantly, Raman spectra of materials or compounds are usually composed of multiple Raman bands. A single spectral band can only be used to calibrate the centered position of a detector, which is not suitable for a broader measuring range. Standard light sources with multiple spectral bands can be used for calibration over a broad measuring range. But it is inconvenient to carry, difficult to operate, requiring annual inspection by a metrological agency and complicated to convert units between wavelength and wavenumber. These problems make it unsuitable for routine laboratory instrument calibration. Therefore, it is necessary to develop CRMs of Raman shift with multi-spectral bands in a broad range for linear calibration of Raman spectroscopy.

National Institute of Metrology, China, has developed five Raman shift CRMs: sulfur, naphthalene, cyclohexane, acetaminophen and polystyrene (GBW13651–GBW13654 and GBW13664), in the form of liquid, bulk solid and powder, for calibration of spectrum within $(83–3,325)$ cm^{-1} of Raman shift. Characteristic values of Raman shift CRMs are determined by a metrological standard of laser confocal Raman spectroscopy ([2015] National Metrological Standard Certificate No. 289), which is traced to the national irradiance metrological standard. Homogeneity and stability of Raman shift CRMs have been investigated, and combined uncertainty and expanded uncertainty have been evaluated. Standard values of characteristic Raman shift are consistent with published data in American standard ASTM E 2911-2013 and domestic interlaboratory measurement results. This indicates that certified values of Raman shift CRMs are reliable, which may provide technical support for R&D application of Raman spectroscopy, reliability and comparability of qualitative and quantitative analysis of Raman spectroscopy and database construction.

3.3.1.1 *Determination method of certified values*

The accuracy of measurement results is determined by measurement method. Generally, measurement conditions mainly affecting Raman spectrum are slit width, lens, grating density, excitation wavelength, accumulation time and laser power. Among them, the changes in

Table 3.2.   Effective measurement conditions for CRMs.[12]

CRM	Laser power/mW	Accumulation time/s
Sulfur	2–4	2–10
Naphthalene	10–15	2–4
Cyclohexane	10–20	2–4
Acetaminophen	10–15	2–4
Polystyrene	10–15	2–4

slit width, lens and grating density have relatively little influence on measurement results; these parameters are generally selected based on experiences and measurement requirements. Accumulation time and laser power have relatively larger influence on Raman spectrum. If laser power is too strong, the tested sample may be burned, therefore, it is necessary to select appropriate laser power to achieve a better signal-to-noise ratio (SNR) without burning the sample. Thermal effect under strong laser exposure may change the position of Raman shift, hence the selection of appropriate accumulation time is also important. If a strong fluorescence signal or relative weak Raman signal is obtained, accumulation time can be increased to obtain better SNR and peak intensity.

An accurate and reliable method of measurement is needed to ensure that all characteristic spectral bands of the tested sample are measured. Therefore, measurement conditions of CRMs have been studied including SNR and laser density at sample surface (laser power and accumulation time). SNR is taken as an important index to evaluate the validity of Raman spectrum measurement program. SNR is calculated through the intensity of a characteristic band with the most sensitive signal response and background noise at the corresponding position. To ensure the integrity of band shape, SNR shall be greater than 100:1 for the most sensitive characteristic band. Finally, laser power and accumulation time are determined when measuring sulfur, naphthalene, cyclohexane, acetaminophen and polystyrene CRMs, as seen in Table 3.2.[12]

3.3.1.2   *Homogeneity test of CRMs*

In accordance with *General Rules and Statistical Principles of Certified Reference Materials* (JJF 1343-2012) (Equivalent ISO

Guide 35:2006), variance analysis method has been used for homogeneity test. Among 100 samples of each type of CRM, a total of 15 bottles of sulfur, naphthalene and acetaminophen, 13 pieces of polystyrene and 10 bottles of cyclohexane are randomly selected, and 3 positions of each selected sample are randomly selected for measurement.

Calculation method of homogeneity test is as follows.[13]

Assuming that m group of samples is extracted, and n times are measured in each group, measurement data are obtained under the same conditions with high repeatability experimental method, and the homogeneity test can be calculated as follows:

$$x_{11}, x_{12}, \ldots, x_{1n1}, \text{ average value } \overline{x_1},$$

$$x_{21}, x_{22}, \ldots, x_{2n2}, \text{ average value } \overline{x_2},$$

$$\ldots \ldots$$

$$x_{m1}, x_{m2}, \ldots, x_m, \text{ average value } \overline{x_m}$$

Total average value

$$\overline{\overline{x}} = \frac{\sum_{i=1}^{m} \overline{x_i}}{m} \tag{3.6}$$

Total number of measurements

$$N = \sum_{i=1}^{m} n_i \tag{3.7}$$

Sum of square deviation between groups

$$Q_1 = \sum_{i=1}^{m} n_i \, \overline{x_i} - \overline{\overline{x}}^2 \tag{3.8}$$

Internal sum of square deviation squares

$$Q_2 = \sum_{i=1}^{m} \sum_{j=1}^{n_i} (x_{ij} - \overline{x_i})^2 \tag{3.9}$$

Degree of freedom (DOF) between groups

$$\nu_1 = m - 1 \tag{3.10}$$

Internal DOF

$$\nu_2 = N - m = \sum_{i=1}^{m} n_i - m \tag{3.11}$$

$$s_1^2 = \frac{Q_1}{\nu_1}, \quad s_2^2 = \frac{Q_2}{\nu_2} \tag{3.12}$$

F distribution variable with degrees of freedom of (ν_1, ν_2)

$$F = \frac{s_1^2}{s_2^2} \tag{3.13}$$

When $s_1 > s_2$, standard deviations of sample difference are

$$s_H^2 = \frac{N(m-1)}{N^2 - \sum_{i=1}^{m} n_i^2}(s_i^2 - s_2^2) \tag{3.14}$$

When $s_1 < s_2$, standard deviations of sample difference are

$$s_H = \sqrt{\frac{s_2^2}{n}} \sqrt[4]{\frac{2}{v_2}} \tag{3.15}$$

According to the degrees of freedom (ν_1, ν_2) and given significance level α, a critical value F_α can be found from the table of F-distribution critical value. If the obtained value F is $F < F_\alpha$ after calculation, no significant difference is found between groups, and the samples are homogenous.

3.3.1.3   *Stability test of CRMs*

Stability of CRMs refers to the ability of a characteristic quantity to remain constant for a given period when CRMs are preserved under certain environmental conditions. Stability includes long-term stability and short-term stability.

Long-term stability refers to the stability of the performance of CRMs under specified storage conditions. Long-term stability is studied by various measurements of characteristic values at different times. According to ISO Guide 35:2006, trend analysis is used for long-term stability test. The first step in evaluating long-term stability is to check whether there is a trend change in the observed data.

Here, the stability is studied using a linear model, which can be expressed by the equation

$$Y = b_0 + b_1 x + \varepsilon \tag{3.16}$$

where x is the time, y is the characteristic value of reference material candidates, b_0 and b_1 are the regression coefficients and ε is the random error component, which can be simply a random error, or include one or more system factors. For stable certified reference materials, the expected value of b_1 is zero.

Assuming that there are n pairs of y observations of x_i, each Y_i can be expressed as follows:

$$Y_i = b_0 + b_1 x_i + \varepsilon_i \tag{3.17}$$

Usually, each Y_i corresponds to a plurality of x_i values due to multiple points of sample extraction at each time point sample. For trend analysis, the average value of all sampling units can be used at time x_i. The regression parameters are therefore calculated as follows:

$$b_1 = \frac{\sum_{i=1}^{n}(x_i - \overline{x})(Y_i - \overline{Y})}{\sum_{i=1}^{n}(x_i - \overline{x})^2} \tag{3.18}$$

$$b_0 = \overline{Y} - b_1 \overline{x} \tag{3.19}$$

Uncertainty of slope b_1

$$s(b_1) = \frac{s}{\sum_{i=1}^{n}(x_i - \overline{x})^2} \tag{3.20}$$

where s is the standard deviation of each point on the line, which can be calculated from the following equation:

$$s^2 = \frac{\sum_{i=1}^{n}(Y_i - b_0 - b_1 x_i)^2}{n - 2} \tag{3.21}$$

Based on b_1, its standard deviation $s(b_1)$ and the distribution factors $t_{(0.95, n-2)}$ with degrees of freedom $n - 2$ and $p = 0.95$ (95% confidence levels), compare $|b_1|$ and $t_{(0.95, n-2)} \cdot s(b_1)$, if $|b_1| < t_{(0.95, n-2)} \cdot s(b_1)$, then the slope is not significant and no instability is observed.

Using trend analysis, characteristic Raman bands of Raman shift CRMs are tested every 3 months within a year. Evaluation results show that the sample has good long-term stability.

Short-term stability refers to the stability of characteristic values of certified reference materials during transportation under prescribed conditions of transportation. It is related to additional impacts arising from the transport of certified reference materials. Simulating current domestic express transportation conditions, samples are tested in the morning and afternoon at 60°C for 3 days. Standard deviation is incorporated as the uncertainty of short-term stability.

3.3.1.4   *Certified value of CRMs*

Under selected measurement conditions, reference material candidates are randomly selected and the measurements are repeated six times independently by two different experimenters. Average value of repeated measurements is taken as the characteristic value of CRMs, and standard deviation is taken as the uncertainty introduced by repeatability.

3.3.1.5   *Uncertainty evaluation of CRMs*

Uncertainty sources include type A uncertainty, type B uncertainty and uncertainty introduced by homogeneity and stability of samples. Type A uncertainty is the uncertainty associated with statistical result of measurement, mainly the uncertainty introduced by measurement process, and standard deviation of repeated measurements is usually used as Type A uncertainty. Type B uncertainty is the uncertainty other than Type A uncertainty, mainly including uncertainty introduced by metrological standard of Raman spectroscopy.

If coverage factor is $k = 2$, then expanded uncertainty U is

$$U = ku_c \tag{3.22}$$

where u_c is combined uncertainty.

Certified values and uncertainties of five kinds of Raman shift CRMs (GBW13651–GBW13654 and GBW13664) are shown in Tables 3.3–3.7.

Table 3.3. Certified value and uncertainty of Raman shift CRMs: Sulfur (GBW13651).

No.	Name	Raman shift/cm^{-1}	Expanded uncertainty $(k = 2)$/cm^{-1}
GBW13651	Raman shift	83.2	2.2
	CRMs: Sulfur	153.2	2.2
		219.2	2.1
		473.2	2.1

Table 3.4. Certified value and uncertainty of Raman shift CRMs: Naphthalene (GBW13652).

No.	Name	Raman shift/ cm^{-1}	Expanded uncertainty $(k = 2)$/cm^{-1}	Raman shift/ cm^{-1}	Expanded uncertainty $(k = 2)$/cm^{-1}
GBW13652	Raman	513.7	2.3	1,381.3	2.2
	shift	763.0	2.2	1,463.5	2.3
	CRMs:	1019.8	2.2	1,576.3	2.2
	Naphthalene	1146.35	2.3	3,055.1	2.3

Table 3.5. Certified value and uncertainty of Raman shift CRMs: Cyclohexane (GBW13653).

No.	Name	Raman shift/ cm^{-1}	Expanded uncertainty $(k = 2)$/cm^{-1}	Raman shift/ cm^{-1}	Expanded uncertainty $(k = 2)$/cm^{-1}
GBW13653	Raman	384.1	2.3	1,444.2	2.2
	shift	426.5	2.4	2,664.2	2.2
	CRMs:	801.9	2.4	2,852.4	2.2
	Cyclohexane	1,028.1	2.2	2,923.4	2.2
		1,157.6	2.4	2,937.5	2.2
		1,266.4	2.3	—	—

3.3.2 Certified Reference Materials of Relative Intensity for Raman Spectroscopy

Rare earth elements doped in borosilicate glass with different components and contents are prepared, and their fluorescence signal collected by Raman spectroscopy has been utilized for relative

Table 3.6.   Certified value and uncertainty of Raman shift CRMs: Acetaminophen (GBW13654).

No.	Name	Raman shift/ cm^{-1}	Expanded uncertainty $(k=2)$/ cm^{-1}	Raman shift/ cm^{-1}	Expanded uncertainty $(k=2)$/ cm^{-1}
GBW13654	Raman	214.1	2.4	1,168.2	2.3
	shift	328.8	2.3	1,236.5	2.3
	CRMs:	391.8	2.2	1,277.7	2.4
	Acetaminophen	465.3	2.2	1,324.4	2.3
		504.3	2.4	1,371.2	2.4
		651.8	2.4	1,515.2	2.3
		710.9	2.2	1,561.4	2.3
		797.1	2.4	1,648.5	2.6
		834.0	2.3	2,930.40	2.2
		857.5	2.2	3,064.6	2.2
		968.5	2.3	3,101.9	2.4
		1,104.6	2.7	3,324.6	2.3

Table 3.7.   Standard value and uncertainty of Raman shift CRMs: Polystyrene (GBW13664).

No.	Name	Raman shift/ cm^{-1}	Expanded uncertainty $(k=2)$/cm^{-1}	Raman shift/ cm^{-1}	Expanded uncertainty $(k=2)$/cm^{-1}
GBW13664	Raman	621.2	2.1	1,583.2	2.2
	shift	795.5	2.2	1,602.4	2.1
	CRMs:	1,001.0	2.1	2,851.3	2.3
	Polystyrene	1,031.2	2.1	2,907.5	2.5
		1,154.6	2.3	3,055.7	2.4
		1,449.0	2.1	—	—

intensity calibration. At present, National Institute of Metrology, China, has released certified reference material of relative intensity for Raman spectroscopy at an excitation wavelength of 514.5 nm (GBW13650) and is now developing a series of certified reference materials of relative intensity for Raman spectroscopy at various excitation wavelengths.

This section takes GBW13650 as an example to describe the determination method of certified reference material, expression of value and process of uncertainty evaluation.

3.3.2.1   *Determination method of standard values*

Through an optical system with a standard light source, instrumental response function R can be measured by Equation (3.5). Equation (3.2) for collected Raman signal (S'_S) of a tested sample can be rewritten as

$$S'_S = P_D \beta'_s s D_s \Omega T A_D Q K t = P_D \beta'_s s D_s \Omega R K t \qquad (3.23)$$

Therefore,

$$\beta'_S D_S = \frac{1}{P_D K t} \times \frac{S'_S}{R} \qquad (3.24)$$

In the equation, D_s and β'_s depend on the sample and are closely related to the actual spectrum that the sample shall produce. P_D is the intensity at which the laser reaches the sample and is a constant. K is the geometric factor which varies with different materials with different optical absorption, but its value is also a constant. t is the accumulation time and is also a constant. Thus, the product of these constants does not affect the shape and relative intensity of Raman bands. Therefore, the actual spectrum of the tested sample can be obtained by normalization of curve S'_S/R using the intensity of its highest band.

Figure 3.11 shows measured and calibrated Raman spectra of GBW13650.[14] As seen from the figure, spectrum of laser glass is continuous and suitable for relative intensity calibration. Calibrated values are used as certified values of relative intensity of CRM.

Since Raman spectrum of laser glass is continuous, it is necessary to carry out homogeneity test, stability test, standard value determination and uncertainty evaluation of corresponding relative intensity at each Raman shift. Detailed measurement results and calculation process can be found in the references.[14]

3.3.2.2   *Homogeneity test of CRMs*

In accordance with *General Rules and Statistical Principles of Certified Reference Materials* (JJF JJF 1343: 2012, ISO Guide 35: 2006) (Equivalent ISO Guide 35:2006), variance analysis method has been used for homogeneity test. Among 100 samples, 20 pieces of reference material candidates are randomly selected, and 3 positions of each

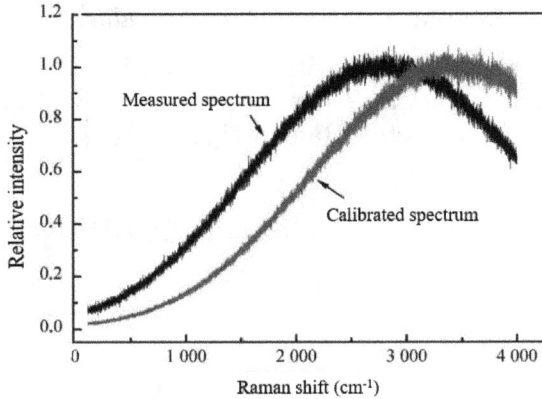

Fig. 3.11.   Measured and calibrated Raman spectra of GBW13650.[14]

selected sample are randomly selected for measurement according to the above-mentioned method. Evaluation results show that the samples are homogenous, and uncertainty introduced by homogeneity is incorporated into uncertainty of CRM.

3.3.2.3   *Stability test of CRMs*

Stability of CRMs is tested by trend analysis method. Raman spectrum of a reference material candidate is measured at three-month intervals over a period of one year. Evaluation results show that the sample is stable, and uncertainty introduced by stability is incorporated into uncertainty of reference material.

3.3.2.4   *Certified value of CRMs*

Under selected measurement conditions, reference material candidates are randomly selected, and measurements are repeated six times independently by two different experimenters. The average of repeated measurements is taken as characteristic value of reference material. Standard deviation is used as uncertainty introduced by repeatability.

3.3.2.5   *Uncertainty evaluation of CRMs*

Uncertainty sources include Type A uncertainty, Type B uncertainty and uncertainty introduced by homogeneity and stability of samples.

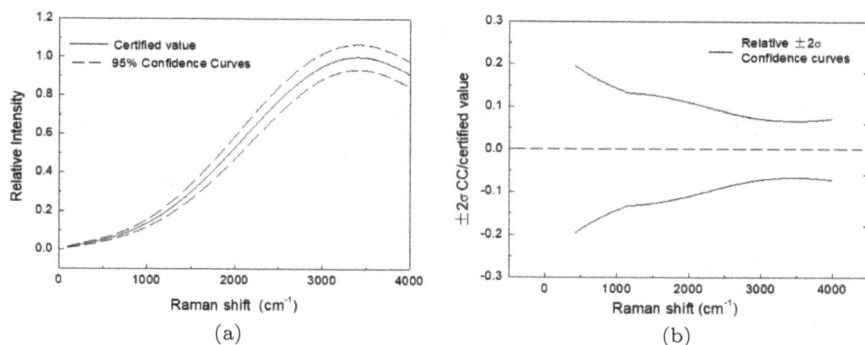

Fig. 3.12. Relative intensity and uncertainty of GBW13650: (a) relative intensity and confidence interval at corresponding Raman shift ($p = 95\%$); (b) relative uncertainty of relative intensity value at corresponding Raman shift ($p = 95\%$).

Type A uncertainty is the uncertainty associated with the statistical result of the measurement, mainly the uncertainty introduced by measurement process, and standard deviation of repeated measurements is usually used as Type A uncertainty. Type B uncertainty is the uncertainty other than Type A uncertainty, mainly including uncertainty introduced by metrological standard of Raman spectroscopy. Please refer to relevant information in Section 3.3.1 on uncertainty evaluation.

Uncertainty evaluation is conducted of relative intensity at each Raman shift. Relative intensity and uncertainty of GBW13650 are shown in Fig. 3.12. Figure 3.12(a) shows relative intensity and confidence interval ($p = 95\%$) for corresponding Raman shift, and Fig. 3.12(b) shows relative uncertainty of relative intensity value at corresponding Raman shift ($p = 95\%$).

To sum up, at present, National Institute of Metrology, China, has released five kinds of Raman shift CRMs, in the form of liquid, bulk solid and powder, suitable for calibration of Raman shift within (83–3,325) cm^{-1}. It has released one kind of CRM of relative intensity for Raman spectroscopy at an excitation wavelength of 514.5 nm (GBW13650). Figure 3.13 shows photographs of these Raman shifts and relative intensity CRMs. Detailed information of CRMs can be found on the website of national certified reference material (www.ncrm.org.cn). CRMs of relative intensity for Raman spectroscopy at excitation wavelengths of 488 nm, 532 nm

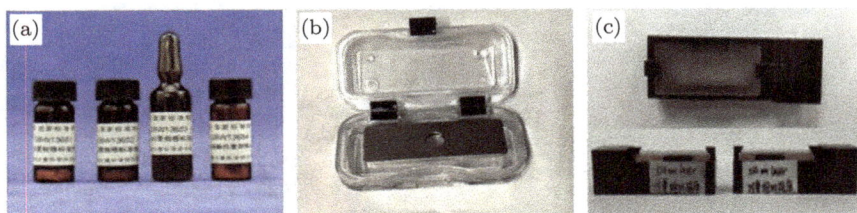

Fig. 3.13. Photographs of Raman spectroscopy-related CRMs: (a) Raman shift CRMs of GBW13651–GBW13654; (b) Raman shift CRM: polystyrene (GBW13664); (c) relative intensity correction standard for Raman spectroscopy: 514.5 nm excitation (GBW13650).

and 785 nm are under development. Raman shift CRMs specially required by biology, medicine, jewelry archaeology and other areas are being developed.

3.4   Calibration Methods of Raman Spectroscopy Based on Raman Shift and Relative Intensity CRMs

For Raman spectroscopy, a daily inspection shall be performed prior to measurement, which can be carried out according to national standard GB/T 33252-2016 *Nanotechnology–Performance testing for laser confocal microscopy Raman spectrometers.*

For Raman spectroscopy calibration, CRMs of Raman shift and relative intensity for Raman spectroscopy are low cost, easy to operate and convenient to carry. These CRMs are necessary for end users to ensure accuracy and reliability of measurement results of Raman spectroscopy in daily use. Calibration procedures include reference material selection, spectrum acquisition, and calibration curve and uncertainty evaluation. At present, relevant national standards and calibration specifications have been issued, such as GB/T 36063-2018 *Nanotechnologies — the Raman shift standard curve for the Raman spectrometer calibration* and JJF 1544-2015 *Calibration specification for Raman spectroscopy.* Calibration standard of relative intensity for Raman spectroscopy is in preparation.

3.4.1   Raman Shift Calibration Based on CRMs

Instrument performance testing is generally carried out using single-crystal silicon, diamond, etc. However, Raman shift of silicon wafer may vary due to surface damage and orientation of silicon wafer. In addition, reference material with a single characteristic Raman band has no practical significance in Raman shift calibration because, for most substances, Raman spectrum covers multiple bands. A single spectral band can only be used to calibrate the centered position of a detector, which is not suitable for a broad measuring range. Therefore, it is necessary to establish a linear calibration method for full range and multiple bands of Raman spectroscopy. Linear calibration through the use of reference materials in accordance with actual measurement range of a specimen is a better and more practical calibration method. The specific process can be found in national standards GB/T36063-2018 *Nanotechnologies — the Raman shift standard curve for the Raman spectrometer calibration*, which describes calibration procedures including reference material selection, spectrum acquisition, calibration curve and uncertainty evaluation. This section introduces these calibration procedures briefly.

In reference materials' selection, Raman shift range of the selected certified reference materials should cover that of the specimen. Under the above conditions, it is advisable to select the reference material which is in a similar solid/liquid state to the specimen.

Take measurement of fullerene by Raman spectroscopy as an example. Raman shift of fullerenes is $(273–1,575)$ cm^{-1}, hence two certified reference materials, sulfur (Raman shift is $(85–474)$ cm^{-1}) and naphthalene (Raman shift is $(513–3,057)$ cm^{-1}), shall be selected. The range of selected Raman shift CRMs covers that of the specimen. Additionally, as fullerene to be tested is a powder, the selected certified reference material sulfur is a bulk solid and naphthalene is a powder. Both selected certified reference materials and the specimen are in solid forms.

In spectrum acquisition, it is advisable to select identical experimental conditions as that of the specimen to be tested, including excitation wavelength, laser power, slit width, wave plate direction and grating. Under selected experimental conditions, the selected sulfur and naphthalene CRMs are measured three times, and average value is taken as measured value of Raman shift CRMs.

In Raman shift calibration curve, measured Raman shift values of reference material are taken as the vertical coordinate, while corresponding standard values are taken as the horizontal coordinate. The obtained data points are fitted linearly. The more fitting points are selected, the better the calibration curve could be drawn, the number

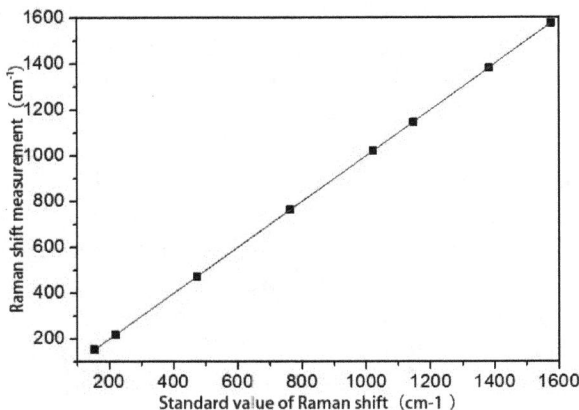

Fig. 3.14.   Raman shift calibration curve and linear fitting using sulfur and naphthalene as certified reference materials.

of fitting points shall be at least five. With average measured values of sulfur and naphthalene taken as the vertical coordinate and corresponding standard values taken as the horizontal coordinate, a graph is drawn as shown in Fig. 3.14. The obtained data points are fitted linearly, and the fitting equation is

$$y = a + bx \qquad (3.25)$$

where y is the measured Raman shift values of the CRMs (cm^{-1}), x is the standard Raman shift values of the CRMs (cm^{-1}), a is the intercept of the fitting line (cm^{-1}) and b is the slope of the fitting line.

The linear fitting result of Fig. 3.14 is

$$y = 1.0005x - 0.5877 \qquad (3.26)$$

After Raman shift calibration using sulfur and naphthalene as certified reference materials, Raman shift of fullerene is measured. Calibrated Raman shift value of fullerene could be obtained by substituting the measured value with Equation (3.26).

Uncertainty introduced by the calibration process includes uncertainty introduced by reference material and uncertainty introduced by calibration curve. Uncertainty u_{plot} introduced by calibration curve $y = a + bx$ is

$$u_{\text{plat}} = \frac{S}{b} \times \sqrt{\frac{1}{p} + \frac{1}{n} + \frac{(\overline{\Delta \nu_m} - \overline{\Delta \nu_{\text{RM}}})^2}{S_{xx}}} \tag{3.27}$$

In the equation,

$$S = \sqrt{\frac{\sum_{i=1}^{n}[\Delta \nu_{m,j} - (a + b\Delta \nu_{c,j})]^2}{n-2}}$$

where S is the residual standard deviation (cm^{-1}), a is the intercept of the fitting line (cm^{-1}), b is the slope of the fitting line, p is the number of measurements of Raman shift of the specimen (number of times), n is the number of measurements of Raman shift of certified reference materials (number of times) and S_{xx} is the quadratic sum of the difference between the measured Raman shifts of all selected CRMs and their average values:

$$S_{xx} = \sum_{j=1}^{n}(\Delta \nu_{c,j} - \overline{\Delta \nu_{\text{RM}}})^2$$

where $\Delta \bar{\nu}_m$ is the average value of measured Raman shift of the specimen (cm^{-1}), $\Delta \bar{\nu}_{\text{RM}}$ is the average value of all Raman shifts measured n times of the CRMs (cm^{-1}), $\Delta \nu_{m,j}$ is the jth measured Raman Shift of CRMs (cm^{-1}) and $\Delta \nu_{c,j}$ is the corresponding jth standard Raman shift of the CRMs (cm^{-1}).

3.4.2 Relative Intensity Calibration of Raman Spectroscopy Based on CRMs

Relative intensity calibration procedures of Raman spectroscopy include reference material selection, spectrum acquisition, calibration curve and uncertainty evaluation. This section introduces these calibration procedures briefly. Detailed process can be found in international and domestic standards: ASTM E2911-2013 *Standard Guide for Relative Intensity Correction of Raman Spectrometers* and T/CSTM 00159-2020 *Calibration Methods of Portable Raman Spectroscopy*. At present, National Institute of Metrology, China, is preparing and applying for relevant national standards.

In reference material selection, it is advisable to select certified reference materials with the same excitation wavelength as the specimen. This time, certified reference materials with excitation wavelengths of 514.5 nm and 785 nm are used.

In spectrum acquisition, it is advisable to select identical or similar experimental conditions as that of the specimen to be tested, including excitation wavelength, laser power, slit width, wave plate direction and grating.

With Raman shift as the horizontal coordinate, normalizing relative intensity of measured spectrum, dividing measured value by standard value of relative intensity, calibration curve R is obtained:

$$R = \frac{I_{\text{RM}}}{I_{\text{SRM}}} \tag{3.28}$$

where I_{SRM} is the standard relative intensity of CRM and dimensionality is 1, I_{RM} is the measured relative intensity of CRM, dimensionality is 1.

Then, Raman spectrum of cyclohexane is measured. Dividing calibration curve from measured spectrum, normalized by maximum intensity at the highest band, calibrated standard Raman spectrum of cyclohexane is obtained. Uncertainty introduced by calibration process includes uncertainty introduced by reference material and uncertainty introduced by the calibration curve:

$$u_{\text{R}} = \sqrt{u_{\text{SRM}}^2 + u_{\text{RM}}^2} \tag{3.29}$$

where u_{SRM} is the uncertainty introduced by CRMs, dimensionality is 1, and u_{RM} is the uncertainty in the measurement of CRMs, dimensionality is 1.

The Raman spectra of cyclohexane before and after Raman's relative intensity calibration are shown in Fig. 3.15. As seen from the figure, Raman spectra without relative intensity calibration are not comparable. Normalize Raman spectra with the intensity at Raman shift of $800 \, \text{cm}^{-1}$ as 1. Before calibration, relative intensity of characteristic band at $2,850 \, \text{cm}^{-1}$ is about 0.7 when using the excitation wavelength of 514.5 nm, while it is 0.2 when using the excitation wavelength of 785 nm, a difference of over 3 times.

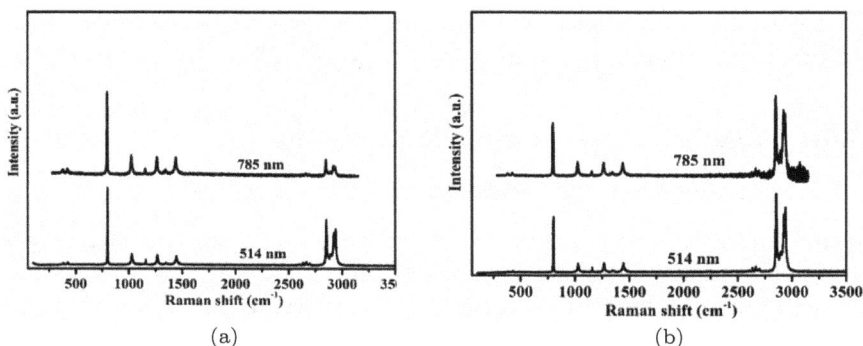

Fig. 3.15. Raman spectra of cyclohexane before and after relative intensity calibration: (a) before calibration; (b) after calibration.

After relative intensity calibration, relative intensities of characteristic Raman peaks at $2,850 \, \text{cm}^{-1}$ are both 1.4 when using the excitation wavelengths of 514.5 nm and 785 nm, and their relative intensity is consistent and comparable.

To sum up, Raman spectroscopy calibration with reference materials can ensure the accuracy and reliability of the measurement results. For end users of Raman spectroscopy in scientific research and industry institutes, it is suggested that reference materials with certificate shall be selected to perform the calibration curve of the characteristic Raman measurand value and standard value. Calibration procedures include reference material selection, spectrum acquisition, calibration curve and uncertainty evaluation.

3.5 Standard Measurement Methods of Graphene-Related Materials by Raman Spectroscopy

Raman spectroscopy is an important characterization method for graphene-related materials. Therefore, accurate measurement methods with data analysis need to be established and recognized as international standards or national standards.

To develop a standardized universal measurement method, the measurement method shall be established, and the feasibility of measurement methods and consistency of measurement results need to be tested by metrological comparison. On this basis, standard shall be prepared and released and is gradually being used by laboratories.

3.5.1   Study on Measurement Methods

3.5.1.1   *Calibration and requirements of instruments*

Raman spectroscopy needs to be calibrated prior to each measurement. This is the premise of accuracy and reliability of measurement. Refer to Section 3.4 for instrument calibration. In addition, for measurement of graphene-related materials, Raman spectroscopy with excitation laser wavelengths of (450–650) nm shall be used, and spectral resolution of the instruments shall be better than $3\,\mathrm{cm}^{-1}$.

3.5.1.2   *Sample preparation*

Raman spectroscopy is a non-destructive analysis technique with the advantage of simple preparation. There is virtually no need for elaborate sample preparations.

Selection of substrates is a key process in measurement of graphene-related materials by Raman spectroscopy. The interaction between substrates and graphene-related materials may cause changes in Raman shift and FWHM of Raman band. According to reported literature,[15–17] Raman shift and FWHM of G band are relatively close for graphene-related materials prepared on SiO_2/Si, quartz, polydimethylsiloxane (PDMS), Si, glass and NiFe substrate, which are $(1581 \pm 1)\,\mathrm{cm}^{-1}$ and $(15.5 \pm 1)\,\mathrm{cm}^{-1}$, respectively. Raman shift and FWHM of G band for graphene-related materials prepared on SiC, indium tin oxide and sapphire substrates are quite different. For single-layer graphene grown by epitaxy on SiC substrate, Raman shifts of G band and 2D band shifted toward higher wave numbers about $11\,\mathrm{cm}^{-1}$ and $34\,\mathrm{cm}^{-1}$, respectively. This is due to the stress effect caused by mismatch between graphene and substrate. There is a layer of carbon atoms with honeycomb lattice structure between graphene and SiC substrate. This lattice mismatch between the carbon layer and the graphene results in compressive stress on graphene causing displacement of Raman shift. For graphene-related materials prepared on sapphire substrate, local density of water molecules varies at the interface of a sapphire substrate with different crystal planes, and blue shift of G band is caused by hole doping of graphene induced by the water layer. On the other hand, red shift mechanism of G band and 2D band of single-layer graphene on an ITO substrate is still not clear. Therefore, for measurement of graphene-related materials by Raman spectroscopy, it is advisable to select quartz, polydimethylsiloxane (PDMS), Si, glass and Si substrates covered by an SiO_2 layer of 300 nm or 90 nm thick.

For graphene-related materials in different forms, sample pretreatment methods are slightly different. Graphene films are transferred to the substrate by solution etching. Graphene powder is placed onto the substrate and then slightly flattened by a clean spoon or slide. Graphene slurry is deposited onto the substrate. If the Raman signal is relatively weak, a vacuum pump or an oven can be used to remove solvent and surfactant residues in the slurry.

3.5.1.3   *Selection of measurement parameters*

In general, Raman shift of the tested sample is independent of excitation wavelength, because Raman shift reflects the energy of lattice vibrational mode, which is an intrinsic property of the material. But for samples with strong fluorescence signals and/or stimulated radiation, choice of different excitation wavelengths has great influence on experimental results. Normally, blue and green lasers are suitable for inorganic materials, resonance Raman experiments (such as carbon nanotubes) and surface-enhanced Raman scattering. Red and near-infrared lasers are suitable for suppressing fluorescence of samples. Ultraviolet laser is suitable for resonance Raman experiment of biomolecules (proteins, DNA, etc.) and suppression of fluorescence of samples. For graphene-related materials, laser wavelengths of (450–650) nm are recommended, including 488 nm, 514.5 nm, 532 nm and 633 nm.

Accumulation time and laser power have great influence on Raman spectrum. The tested sample may be burned under too powerful laser, hence selection principle of laser power is to achieve a better signal-to-noise ratio without burning the sample. If laser is too strong, thermal effect may cause changes of Raman shifts, hence it is important to choose appropriate accumulation time. For graphene-related materials, it is advisable that the laser power at the sample surface be less than 5 mW, and the total accumulation time be less than 60 s.

Grating density and slit width have great influence on signal intensity and resolution of Raman spectrum. Higher grating density reflects stronger dispersion ability; therefore, selection of higher-density grating can improve spectral resolution. However, higher grating density may decrease the single measurement range and efficiency of Raman spectrum. Wider slit results in stronger Raman signal intensity but worse resolution. Conversely, smaller slit results in weaker Raman signal intensity but better resolution. Therefore, for graphene-related materials, it is necessary to select appropriate grating density and slit width for balanced signal intensity and resolution.

Scanning range of $(100–3,100)\,\mathrm{cm}^{-1}$ is recommended for graphene-related materials, which covers characteristic Raman bands of graphene including D band, G band and 2D band (G$'$ band). For sp^2 carbon materials, besides these bands, there are C band caused by interlayer shear mode (usually appearing at $(25–50)\,\mathrm{cm}^{-1}$), respiration mode induced by interlayer breath vibration (usually studying its harmonic band positioned at $(145–220)\,\mathrm{cm}^{-1}$) and other second-order sum and double frequency Raman bands $((1{,}650–2{,}300)\,\mathrm{cm}^{-1})$. These Raman signals are easy to be ignored because of their weak intensities, but Raman shift, band shape and intensity of these Raman characteristic bands are strongly dependent on number of graphene layers and stacking method between layers. With the development of materials and equipment, for Raman spectroscopy with excellent performance such as low wave number and high resolution, scanning range can be adjusted appropriately. Through analysis of Raman spectra of these weak signals, we can further study electron–electron and electron–phonon interactions, and their Raman scattering processes in graphene-related materials.

In terms of sampling, various positions of samples shall be measured. For graphene films, it is advisable to measure at least six positions including center and edge of the sample. For graphene powders and slurries, it is advisable to take at least six samplings from top, middle and bottom of the sample bottle.

3.5.1.4   *Data processing*

For measured spectra, background noise caused by detector, thermal charge and other interference factors can be deducted by dark correction or dark subtraction. Lorentz or Gaussian fitting is used to determine Raman shift, FWHM and intensity of characteristic bands.

3.5.1.5 *Uncertainty evaluation*

Uncertainty sources of measurement results include uncertainty introduced by equipment calibration and uncertainty introduced by measurement method. Uncertainty introduced by equipment calibration u_1 is evaluated according to Sections 3.4.1 and 3.4.2 for Raman shift and relative intensity, respectively. The uncertainty introduced by measurement method includes uncertainty introduced by repeatability u_2 and the uncertainty introduced by sampling u_3 (standard deviation). The combined uncertainty u_c is

$$u_c = \sqrt{u_1^2 + u_2^2 + u_3^2} \tag{3.30}$$

expanded uncertainty U with confidence of 95% and coverage factor k of 2 is

$$U = k^* u_c (k = 2) \tag{3.31}$$

3.5.2 Domestic Comparison for the Standardization of Measurement Methods

To verify universality of measurement methods and comparability of measurement results, National Institute of Metrology, China, led the domestic comparison of *Measurement of graphene-related materials by Raman spectroscopy*. Based on the study in Section 3.5.1, the lead laboratory prepared comparison protocol and samples, collected measurement results performed by participating laboratories, conducted statistical analysis of measurement results and analyzed sources of measurement uncertainty. In the domestic comparison, the lead laboratory provided a reference material for Raman shift, which was used for instrument calibration, and two kinds of graphene-related materials, Sample A and Sample B, for measurement. Raman spectra measurement of the samples and data analysis were performed according to the procedures described in the protocol.

The lead laboratory analyzed measurement results performed by participating laboratories. Uncertainty of measurement results of participating laboratories was calculated according to

Equations (3.30) and (3.31). Consistency of measurement results was evaluated by using normalized deviation E_n statistical method. According to JJF1117-2010 *Metrological Comparison*, E_n is calculated as

$$E_n = \frac{Y_{ji} - Y_{ri}}{k \times u_i} \qquad (3.32)$$

In the equation, k is the coverage factor, generally $k = 2$, and u_i is the standard uncertainty of $Y_{ji} - Y_{ri}$ at the ith measurement.

When u_{ri}, u_{ji} and u_{ei} are unrelated or weakly correlated,

$$u_i = \sqrt{u_{ri}^2 + u_{ji}^2 + u_{ei}^2} \qquad (3.33)$$

In the equation, u_{ri} is the standard uncertainty of reference values at the ith measurement, u_{ji} is the standard uncertainty of measurement results at the ith measurement in the j laboratory and u_{ei} is the effect of instability of transfer standard on measurement results at the ith measurement.

If $|E_n| \leq 1$, the difference between measurement results from the participating laboratory and the reference value is within a reasonable expectation, and the comparison results are acceptable.

Take Sample A, for example, Raman shift measurement results of participating laboratories are shown in Table 3.8 and Fig. 3.16. It can be seen that uncertainty levels of measurement results from

Table 3.8.   Raman shift measurement results of Samples A.

Number of participating laboratory	Raman shift of D band/ cm^{-1}	Expanded uncertainty $(k = 2)$/ cm^{-1}	E_n	Raman shift of G band/ cm^{-1}	Expanded uncertainty $(k = 2)$/ cm^{-1}	E_n
1	1,338	12	−0.24	1,599	19	0.37
2	1,341	11	0.09	1,591	20	−0.06
3	1,334	14	−0.47	1,579	19	−0.69
4	1,343	11	0.24	1,599	18	0.36
5	1,348	20	0.41	1,588	20	−0.22
6	1,346	10	0.50	1,600	18	0.45
7	1,337	12	−0.28	1,589	18	−0.19
8	1,341	16	0.04	1,602	19	0.49
9	1,340	10	−0.01	1,586	19	−0.37

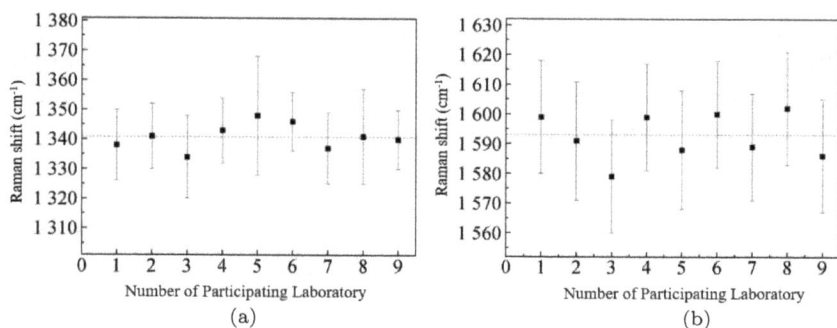

Fig. 3.16. Raman shift measurement results of Samples A: (a) D band; (b) G band.

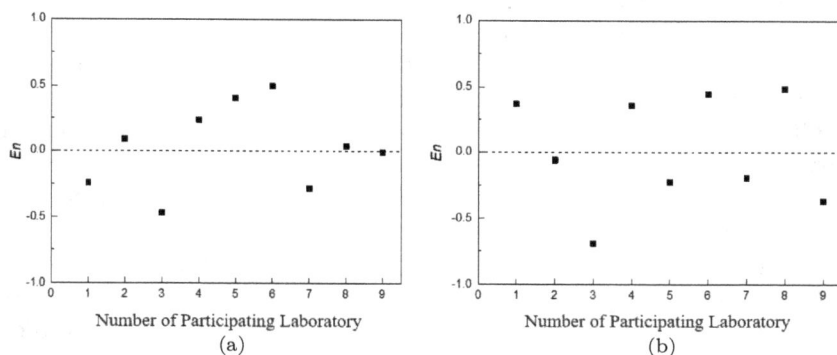

Fig. 3.17. E_n value of Samples A: (a) D band; (b) G band.

various laboratories are similar, therefore, the arithmetic mean of measurement results from nine laboratories was used as reference values.

Figure 3.17 shows E_n values of Samples A. From analysis results, $|E_n|$ values from all participating laboratories are less than 1, indicating that the differences between the measurement results and the reference value are within reasonable expectation. The analysis results show that the measurement method is reliable, operable and universal, which can ensure consistency of measurement results.

Based on the above measurement methods and domestic comparison, National Institute of Metrology, China, has prepared and published an association standard T/CSTM 00166.1-2019 *Characterization of Graphene Related Materials — Part I: Raman*

Spectroscopy. Together with the measurement methods by X-ray diffraction, atomic force microscope and transmission electron microscope, a series of standards could be used for the identification of graphene-related materials.

3.6　Summary

Raman spectroscopy is an effective technique to characterize the structural and characteristic parameters of graphene-related materials, such as layers, defects, doping and stacking. Therefore, accurate measurements of Raman spectra are essential for preparation, development and production of graphene-related materials in both scientific and industrial areas.

Traceability is the guarantee of accurate and reliable measurement results. For end users, instrument calibration through the use of CRMs is convenient and practical, and traceability of CRMs can ensure accuracy and reliability of measurement results. The most important issue in instrument calibration with certified reference materials is to select the reference materials close to the tested samples. Take Raman shift calibration for example, there are many kinds of CRMs issued at present, and the CRMs need to be selected based on the wavenumber range and form (solid or liquid) of tested samples.

Measurement results of the characteristic parameter of material depend on the status of the material itself and the measurement process, therefore, standard measurement methods play an important role in material metrology. This chapter has introduced sample preparation, measurement parameters selection, spectrum and data analysis, and uncertainty evaluation processes for the measurement of graphene-related materials by Raman spectroscopy. To ensure the universality of measurement methods and reliability of measurement results, NIM, China, organized a related domestic comparison, which may facilitate the development and adoption of national standards.

Raman spectroscopy is only one of the methods for graphene-related materials' characterization. The crystallinity and microstructure of graphene-related materials prepared by different methods may be different. In practice, a variety of methods have to be utilized together to analyze characteristics of graphene-related materials. Together with measurement results by X-ray diffraction, atomic force microscope and transmission electron microscope, measurement results by Raman spectroscopy can be used as one criterion for the identification of graphene-related materials, providing technical guidance for research and development of graphene-related materials.

References

[1] Wu J B, Lin M L, Cong X, *et al.* Raman spectroscopy of graphene-based materials and its applications in related devices. *Chemical Society Reviews*, 2018, 47(5): 1822–1873.

[2] Yao Y X, Ren L L, Gao S T, *et al.* Histogram method for reliable thickness measurements of graphene films using Atomic Force Microscope (AFM). *Journal of Materials Science & Technology*, 2017, 33(8): 815–820.

[3] Moon I K, Lee J, Ruoff R S, *et al.* Reduced graphene oxide by chemical graphitization. *Nature Communications*, 2010, 1: 73.

[4] Gupta A, Chen G, Joshi P, *et al.* Raman scattering from high frequency phonons in supported n-Graphene layer films. *Nano Letters*, 2006, 6(12): 2667–2673.

[5] Sheng, Xiangyong, Ren, Lingling, Yao, Yaxuan, *et al.* Effect of temperature on interlayer coupling of stacked double-layer graphene. *Journal of Metrology,* 2018, 39(6): 791–796.

[6] Wu, Juanxia, Xu, Hua, Zhang, Jin. Application of Raman spectroscopy in the characterization of graphene structure. *Journal of Chemistry*, 2014, 72(3): 301–318.

[7] Polard A J, Brennan B, Stec H, *et al.* Quantitative characterization of defect size in graphene using Raman spectroscopy. *Applied Physics Letters*, 2014, 105(25): 253107.

[8] Zhao, Yingchun, Ren, Lingling, Wei, Weisheng, *et al.* Calibration procedure for laser confocal micro-Raman Spectrometer. *Spectroscopy and Spectral Analysis*, 2015, 35(9): 2544–2547.

[9] Fryling M, Frank C J, Mcrery R L. Intensity calibration and sensitivity comparisons for CCD/Raman spectrometers. *Applied Spectroscopy*, 1993, 47(12): 1965–1974.

[10] Yao, Yaxuan, Ren, Lingling, Gao, Sitian, *et al.* Calibration of Raman Spectrometer relative intensity and evaluation of uncertainty. *Modern Scientific Instruments*, 2015, (3): 134–139.

[11] National Committee for the management of certified reference materials. *Development, Management and Application of Certified Reference Materials*. Beijing: China Metrology Publishing House, 2010.

[12]   Ren, Lingling, Zhao, Yingchun, Yao, Yaxuan, *et al.* Determination of effective measuring procedures for Raman Spectra of several representative pure substances. *Modern Measurement and Laboratory Management*, 2014, (6): 3–6.

[13]   Quan, Hao, Han, Yongzhi. *Certified Reference Materials and Their Application Technology* (2nd edn.). Beijing: China Standard Press, 2003.

[14]   Yao, Yaxuan, Ren, Lingling, Gao, Sitian, *et al.* Expression of mass value of relative intensity reference material of Raman Spectrometer and evaluation of uncertainty. *Journal of Metrology*, 2017, 38(3): 376–379.

[15]   Wang Y Y, Ni Z H, Yu T, *et al.* Raman studies of mono layer graphene: The substrate effect. *The Journal of Physical Chemistry C*, 2008, 112(29): 10637–10640.

[16]   Das A, Chakraborty B, Sood A K. Raman spectroscopy of graphene on different substrates and influence of defects. *Bulletin of Materials Science*, 2008, 31(3): 579–584.

[17]   Komurasaki H, Tsukamoto T, Yamazaki K, *et al.* Layered structures of interfacial water and their effects on Raman spectra in graphene-on-sapphire systems. *The Journal of Physical Chemistry C*, 2012, 116(18): 10084–10089.

https://doi.org/10.1142/9789811295713_0004

Chapter 4

Measurement of the Crystal Structure of Graphene-Related Materials by X-ray Diffraction

Huifang Gao

4.1 Overview

X-ray diffraction (XRD) technique is one of the main methods to characterize the crystal structure of materials, which has the advantages of low sample consumption, simple preparation and non-destructiveness. As such, it has extensive and important application in chemistry, physics, geology, materials science, biology and other disciplines, in petroleum, chemical, metallurgical, information industry, aerospace and other industrial sectors, as well as the fields of judicature, commodity appraisal and so on. The XRD technique utilizes the diffraction effect of X-rays in crystals and non-crystals to characterize the crystal structure of the material, interplanar distance, crystal lattice parameters and crystallinity.[1]

Graphene-related materials include graphene oxide (GO) with varying degrees of oxidation or reduction, reduced GO and at least a single layer of graphene. To distinguish these materials, XRD is an important tool to characterize the crystal structure of graphene-related materials. XRD can measure the diffraction peaks of different graphene materials, or calculate layer distance by Bragg's law. Combined with other methods, it can be used as a basis for measuring the crystal structure of graphene materials; its accurate measurement can provide technical guidance for the production and research of graphene materials.

Since the differences in the number of holes, defects, water molecules and –COOH, –COH, –OH groups contained in the layers make the interlayer distance d_{002} different, the angle 2θ of the GO corresponding to the XRD spectrogram is different. Huh[2] reported the XRD figures of GO and graphene in the thermal reduction process of GO (Figs. 4.1 and 4.2) and gave the schematic figure for measuring the structure of the graphenematerial (Fig. 4.3). It is clear from Figs. 4.1 and 4.2 that the higher the degree of oxidation of graphene, d_{002} corresponding peak 2θ appears at a smaller angle of $22°$ $-25°$, and the wetter the GO sample, the weaker the inner peak intensity, or even hardly appear. See Figs. 4.2(a)–(c) and 4.3(e).

With the increase in reduction degree and the loss of water, d_{002} corresponding peak 2θ moves to a large angle (right), and there are two distinct peaks within $13°$–$25°$. As the reduction becomes more and more complete, d_{002} corresponding peak 2θ of graphene is almost identical to the graphite 2θ, but the peak width of graphene is larger than that of graphite. This reflects that the thermal reduction process of GO includes the removal of embedded water molecules and oxygen-containing functional groups, defect formation of GO/Graphene sheet, lattice shrinkage and stripping, folding and unfolding of GO/Graphene layers, and the bottom-up trend of bulk graphite layer by layer.

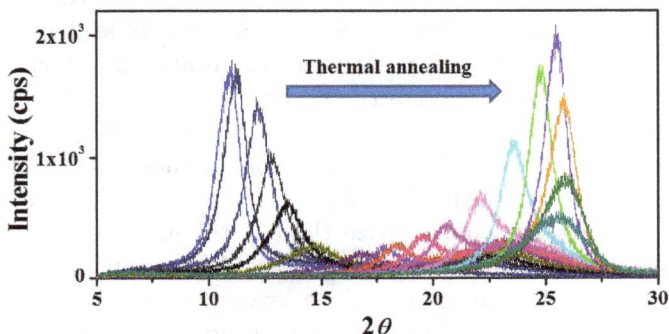

Fig. 4.1.   Total plot for measured *in situ* XRD patterns of GO/GP films obtained at RT to $1,000°$C.

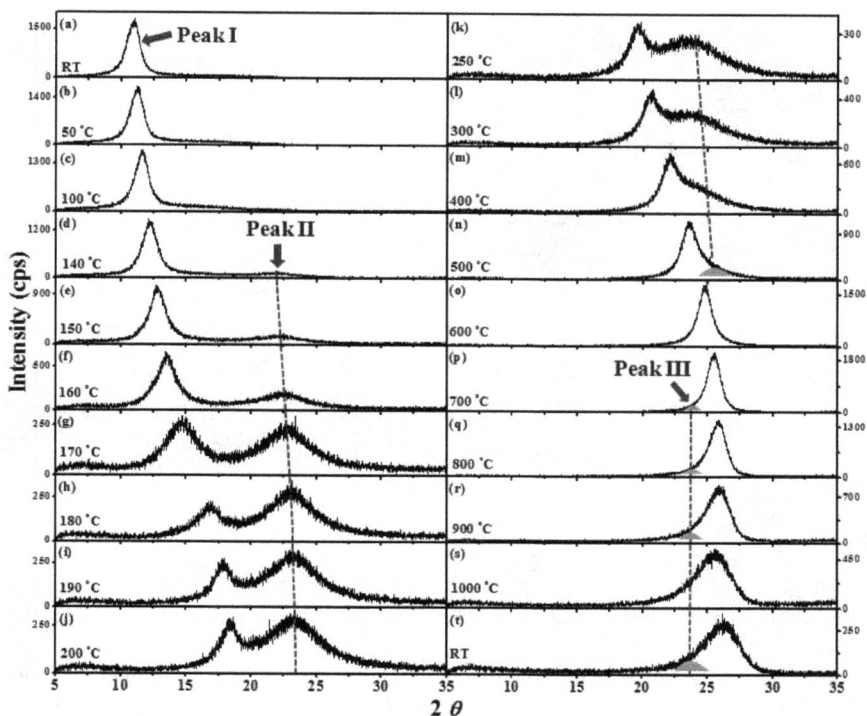

Fig. 4.2.   Individual XRD plot for Fig. 4.1.

To better compare the XRD graphical differences between graphite and graphene in terms of oxidation and reduction of oxides, Fig. 4.4 gives the XRD plots of Graphite Oxide (GOt) and reduction of GOt by reducing agents with different reducibility.[3] A comparison of Figs. 4.4 and 4.2(a)–(c) shows that at room temperature the difference between GOt and GO is that GOt has peaks within 20°–25° while GO has no peaks. Through reduction in an ethanol solution, on the one hand, different degrees of reduction are conducted; on the other hand, graphite stripping into a few layers of graphene is promoted. Thus, in the presence of methanol (MeOH), ethanol (EtOH), isopropanol (iPrOH), benzyl (BnOH) and hydrazine reductants, GOt is reduced to a few layers of graphene.

It can be seen from the above analysis that the peak angle 2θ is an important parameter for the characterization of graphene, which

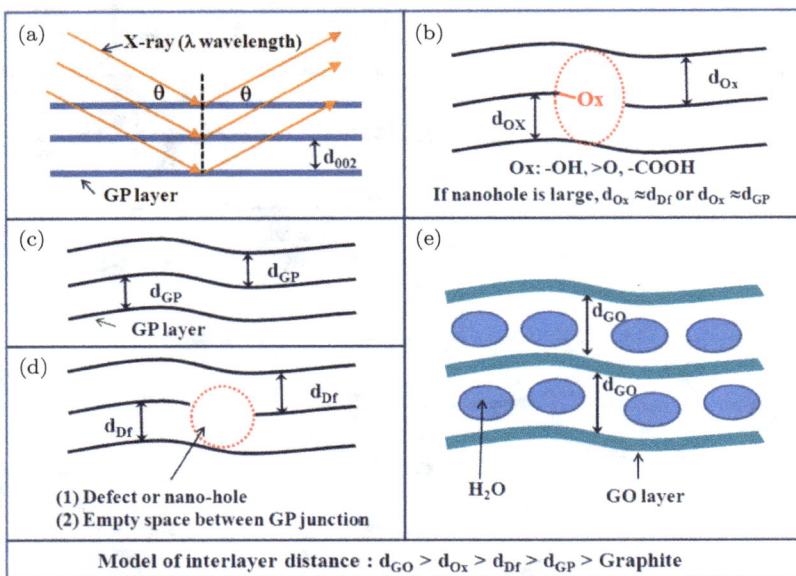

Fig. 4.3.   Bragg's law for GP or graphite (002) planes (a); models for d_{002} (e); thermally reduced GP (b–d).

Notes: d_{Gt} — the interlayer distance of graphite; d_{GP} — the interlayer distance of GP; d_{Df} — the interlayer distance of thermally reduced graphene containing defects and nano-holes; d_{Ox} — the interlayer distance of GO or GP containing oxide groups of C–O$_x$ with an sp^3 bond; d_{GO} — the interlayer distance of GO containing intercalated H$_2$O molecules and various oxygen groups.

is vital to analyzing the oxidation state of graphene, the interlayer distance and the difference between graphene and graphite, hence accurate measurement is required. To achieve accurate measurement, it is necessary to ensure five aspects: the human (staff), machine (equipment), materials (samples), law (regulation, standard) and environment. In terms of the XRD method for measuring the crystal structure of graphene materials, assuming the personnel and environment meet the requirements, the subject graphene material has been selected; the accuracy of its crystal structure by XRD method is related to whether the XRD equipment is accurate and reliable and whether the method is reliable and consistent. Therefore, this chapter focuses on how to ensure the accuracy and reliability of the XRD equipment, and the reliability and consistency of the crystal structure of graphene-related materials by XRD method.

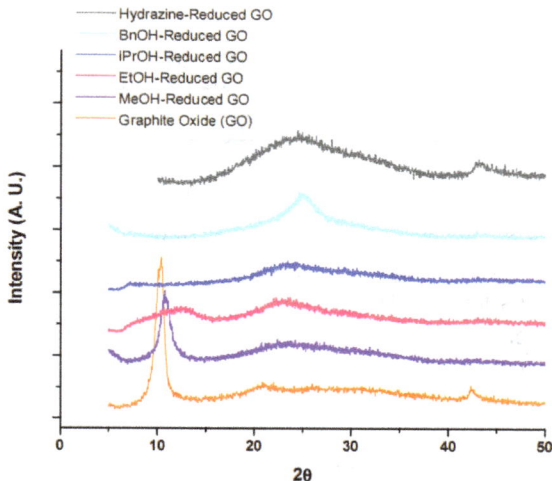

Fig. 4.4. XRD plots of graphite oxide and reduction of graphite oxide by MeOH, EtOH, iPrOH, BnOH and hydrazine (experiment condition $\lambda = 0.15405$ nm, temperature is 25°C, relative humidity is 45%).

4.2 Equipment Traceability and Calibration

As is mentioned in the last section, in order to measure the crystal structure of graphene by XRD technique, it is necessary to ensure the main performance of the instrument, particularly the accuracy and reliability of angle measurement results, so that the needs of graphene quality control can be met in the development of graphene. To ensure the accuracy and reliability of quantity value, the most basic and core process is the value tracing and transfer of the instrument. Traceability is defined as a continuous chain of comparison with specified uncertainty so that the measurement result or the value of the measurement standard can be compared with the specified reference standard, usually a characteristic associated with a national or international standard of measurement. The term "dissemination of the value of quantity" refers to the dissemination of values through metrological verification or calibration, an activity in which the measurement unit values reproduced from the national measurement standards are gradually disseminated to the measurement standards at different levels until the work of measuring devices. Traceability and dissemination of quantity value are the

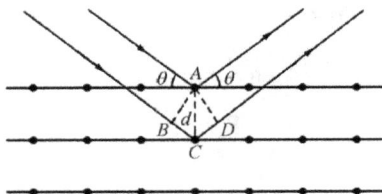

Fig. 4.5.   Schematic figure of X-ray diffraction technique.

technical studies that national institutes of metrology need to carry out; the national metrological standard devices, certified reference materials and technical specifications for calibration are established through research to provide to the social end users (scientific research institutions, enterprises, etc.).

XRD technique utilizes the diffraction effect of crystals on X-ray; the internal structure of the crystal is identified by the diffraction characteristics of the X-ray as it passes through the crystal lattice of the material. This method is based on Bragg's law, and the equation can be given by Equation (4.1). The schematic figure is shown in Fig. 4.5:

$$2d_{(hkl)} \sin \theta = n\lambda \qquad (4.1)$$

where $d_{(hkl)}$ is the interplane distance (nm), θ is the diffraction angle (°), n is the order of the diffraction and λ is the wavelength of X-ray (nm).

Among them, h, k, l represent the Miller indices and λ depends on the cathode (target) metal material used in the X-ray tube.

4.2.1   Device Traceability

It can be seen from Equation (4.1) that the accurate measurement of interplane distance is related to the accurate measurement of diffraction angle θ and incident wavelength λ, so it is necessary to trace these two parameters back to the SI base units. The traceability study of the X-ray diffractometer is shown in our previous research papers[4,5]; the sources and values of uncertainty that affect the accuracy of the results of the whole traceability evaluation are shown in Table 4.1. Based on these studies, National Institute of Metrology, China (hereinafter referred to as NIM), has established the tracing path of the X-ray diffractometer (Fig. 4.6(a)), diffraction angle is

Table 4.1.   Evaluation results of sources of uncertainty and quantities for X-ray diffractometer.

Uncertainty component u_i		Uncertainty source	Uncertainty $u(x_i)$	c_i		Coverage factor k	Degree of freedom v
				Symbol	Value		
$u_{\theta 1}$	Angle	Accuracy of angle measurement system	$0.0005°$	c_1	$c(\theta) - 2.518$	2	50
$u_{\theta 2}$		Light paths and detectors	$0.003°$	c_2		2	12.5
$u_{\theta c}$		Total angle of reflection	$0.05°$	c_3	$c(\theta_c)$ 0.064	2	12.5
u_λ		X-ray Wavelength	1.4×10^{-4} nm	c_4	$c(\lambda)$ 5.744	2	50
		Uncertainty u_c					0.01 nm

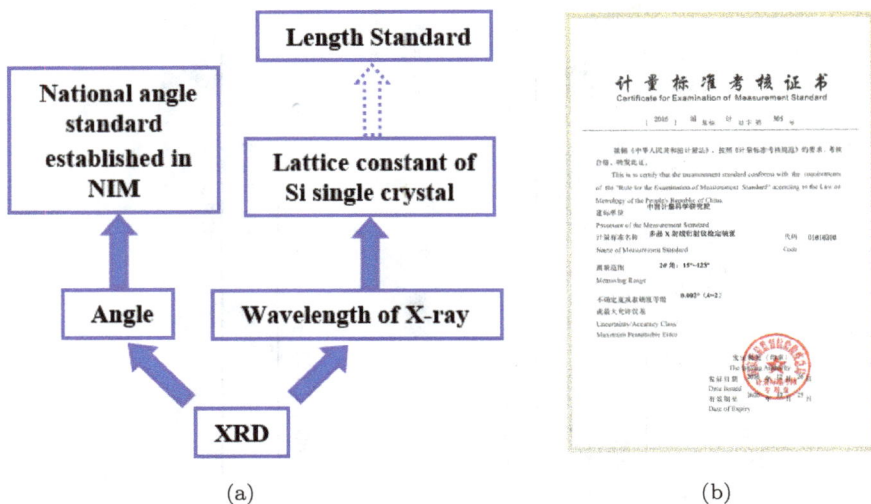

Fig. 4.6.   (a) Tracing path of X-ray diffractometer; (b) calibration equipment certificate of polycrystalline X-ray diffractometer.

traced to the the national angle standard established in NIM, and the incident wavelength is traced to the lattice constant of Si single crystal. At the same time, it has established a calibration device of polycrystalline X-ray diffractometer ((2016) National Standard Metrological Certificate No. 305) for the calibration of the X-ray diffractometer, as shown in Fig. 4.6(b).

4.2.2   Equipment Calibration

End users are more concerned about how to use the results of measurement standards to ensure the accuracy and reliability of the equipment used in the laboratory. There are two main approaches to ensure the continued reliability of equipment. One is that a registered metrological inspection and calibration department carries out regular verification and calibration of the equipment. The other is that the laboratory uses certified reference materials to carry out self-calibration or inspection of the equipment on a regular basis. Both approaches need to be based on standards or regulations.

In accordance with *Polycrystalline X-ray Diffractometer*, the national metrological verification regulation JJG 629-2014, we describe the process of calibrating the X-ray diffractometer as follows. Calibration items mainly include instrument 2θ angle indication error and repeatability, instrument resolution, energy spectrum resolution of detector and stability of diffraction intensity. The calibration items need to select different certified reference materials for calibration, among which the certified reference material for calibration of instrument 2θ angle and instrument resolution is powder α-SiO$_2$ (average particle size is not more than $20\,\mu$m and the standard uncertainty of the lattice constant is not greater than 0.00001 nm). The certified reference material for calibration of energy spectrum resolution and stability of diffraction intensity is powder Si (average particle size is about $10\,\mu$m and the standard uncertainty of lattice constant is not more than 0.00002 nm).

4.2.2.1 *2θ angular indication error of instrument*

Measure certified reference materials of powder α-SiO$_2$ under the following conditions: CuK_α radiation, Ni filter plate, the divergence slit and the scattering slit are set at $1°$, the receiving slit is $(0.1–0.3)$ mm, the continuous scan speed is not greater than $2°$/min, the step scan speed is not greater than $0.01°$/step and unidirectional scan within is $15°$–$125°$ (2θ). Record the diffraction angles of crystal plane (100) (101) (110) (200) (211) (312) and (314) of α-SiO$_2$ reference material and calculate the error between the indication value and the reference value of each diffraction angle according to Equation:

$$\Delta(2\theta) = 2\theta - 2\theta_s \qquad (4.2)$$

where $\Delta(2\theta)$ is the indication error of 2θ angle ($°$), 2θ is the instrument indication of 2θ angle ($°$) and $2\theta_s$ is the 2θ angle corresponding to each crystal plane of the reference material ($°$).

The indication errors' value corresponding to the maximum absolute value among diffraction angle indication errors is the result of the indication error of instrument 2θ angle, which is required to be within $\pm 0.02°$.

4.2.2.2 *Repeatability of instrument 2θ angle*

According to the above measurement conditions, scan unidirectionally the 2θ angle of the crystal plane (101) of the powder α-SiO$_2$

reference material. Repeat it seven times, and calculate the standard deviation based on Equation (4.3):

$$s(2\theta) = \sqrt{\frac{\sum_{i=1}^{n}(2\theta_i - \overline{2\theta})^2}{n-1}} \qquad (4.3)$$

where $s(2\theta)$ is the standard deviation of 2θ angle measurement, 2θ is a single measurement value of 2θ angle (°), $\overline{2\theta}$ is the measurement average of 2θ angle (°) and n is the number of measurements.

Instrument 2θ angle repeatability is required to not exceed 0.002°.

4.2.2.3 Instrument resolution

With reference to the above measurement conditions, the receiving slit is (0.10–0.15) mm, continuous scan speed is not greater than 2°/min and step scan speed is not greater than 0.01°/step. Scan 2θ angle at 67°–69° and record it, then obtain the diffraction pattern shown in Fig. 4.7.

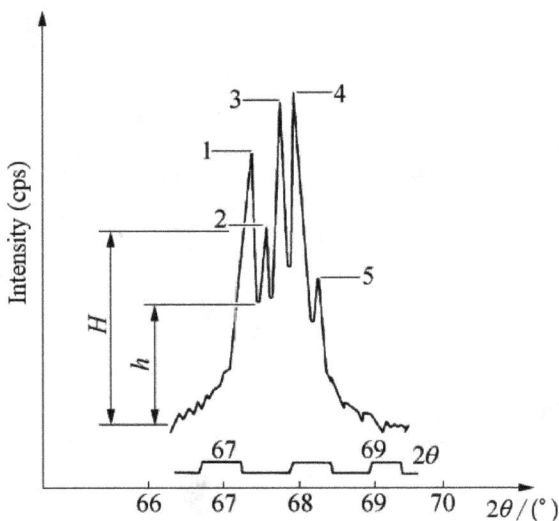

Fig. 4.7. The diffraction pattern of powder α-SiO$_2$ reference material.
Notes: 1 — $K_{\alpha 1}$ diffraction peak of the (212) crystal plane; 2 — $K_{\alpha 2}$ diffraction peak of the (212) crystal plane; 3 — $K_{\alpha 1}$ diffraction peak of the (203) crystal plane; 4 — $K_{\alpha 2}$ diffraction peaks of the (203) crystal plane and $K_{\alpha 1}$ diffraction peak of the (301) crystal plane; 5 — $K_{\alpha 2}$ diffraction peaks of the (301) crystal plane.

Calculate the instrument resolution according to Equation (4.4):

$$D = \frac{h}{H} \times 100\% \tag{4.4}$$

where D is the instrument resolution, H is the diffraction intensity corresponding to the peak-valley position between $K_{\alpha 1}$ and $K_{\alpha 2}$ diffraction peak of the (212) crystal plane and H is the diffraction intensity corresponding to the peak height of the $K_{\alpha 2}$ diffraction peak of the (212) crystal plane.

The instrument resolution is required to be not greater than 60%.

4.2.2.4   *Energy spectrum resolution of the detector*

Measure the reference material of powder Si under the following conditions: CuK_{α} radiation, Ni filter plate, the divergence slit and the scattering slit are set at 1°, and the receiving slit is (0.1–0.3) mm. Adjust $K_{\alpha 1}$ diffraction intensity of Si(111) crystal plane to about 80% of the full value, fix the channel width of the amplifier, and start scanning the energy spectrum resolution, then obtain the scan of energy spectrum resolution, as shown in Fig. 4.8.

Calculate the energy spectrum resolution of the detector according to Equation (4.5):

$$E_R = \frac{W}{V} \times 100\% \tag{4.5}$$

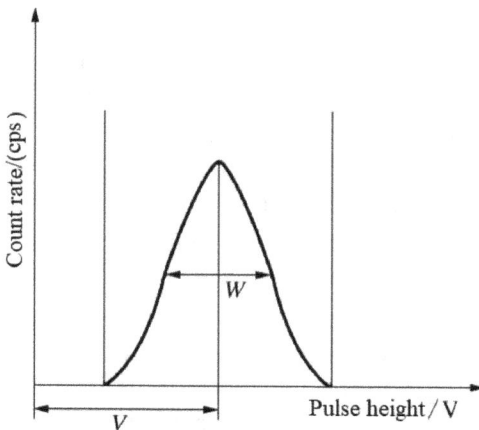

Fig. 4.8.   Scan of energy spectrum resolution of powder Si reference material.

where E_R is the energy spectrum resolution of the detector, W is the half-peak width of the scanning curve, i.e. the width corresponding to the position of half the peak height (V), and V is the voltage corresponding to the highest peak of the scan curve (V).

The energy spectrum resolution of the detector, for proportional counter, is not greater than 20% (CuK_α), and for scintillation counter, it is less than 55% (CuK_α).

4.2.2.5 *Stability of diffraction intensity*

When the newly purchased X-ray diffractometer is first calibrated, or when the installation and maintenance of the instrument has a significant impact on the metrological performance at subsequent calibrations, it is necessary to confirm the stability of diffraction intensity. Measure the reference material of powder Si under the following conditions: Cu K_α radiation, Ni filter plate and the divergence slit are $2°$, the scattering slit is at $4°$, and the receiving slit is over 0.3 mm. Keep the diffraction angle stable and determine the diffraction intensity of the Si(111) crystal plane. Using the timing counting method, the accumulative counting rate is about 1×10^4 times per second, and the timing time is 200 s. After the instrument is settled, record the count every 5 minutes for 8 hours in a row, and calculate the relative range of this set of data according to Equation (4.6):

$$R = \frac{N_{\max} - N_{\min}}{\overline{N}} \times 100\% \qquad (4.6)$$

where R is the relative range of diffraction intensity, $\overline{N}$ is the average diffraction intensity, $N_{\max}$ is the maximum diffraction intensity and $N_{\min}$ is the minimum diffraction intensity.

The stability of diffraction intensity shall not be greater than $1.5\%/8$ h.

According to the above method, the instrument is calibrated and checked regularly to ensure the use of equipment in the quality operation system control.

4.3   Research on Measurement Methods

Section 4.2 describes how to ensure the accuracy and reliability of XRD equipment through tracing and calibration. On the basis of reliable equipment, this section introduces the establishment of the standard method for XRD measuring the crystal structure of graphene powder and related materials.[6]

4.3.1 Preparation of Measurement Samples

Usually, for the crystal structure characterization of powder materials, the measurement method of X-ray powder diffraction is simple and straightforward, but poor sample preparation will cause a fairly big difference in the measurement diffraction angle. Sample preparation is difficult because of the characteristics and different preparation method of graphene powder and related materials. Graphene powder and related materials have low density and are lightweight, which makes it easy to float in the preparation of XRD-measured samples. Their two-dimensional lamellar structure makes the samples easy to slip but hard to compact during preparation, and the sample surface is not smooth, so researchers have studied the sample preparation technology.

By experimenting with different tablet pressing techniques, the researchers have found a solution to the problem of sample floating and slippage. It is recommended that the graphene powder samples be placed in the groove of a sample carrier and the surface of the slide be coated with high-purity ethanol and pressed hard enough until the surface of the sample is in the same plane as that of the sample carrier.

4.3.2 Sampling Principle

To ensure the reliability and representativeness of the measurement results, it is necessary to take samples from different parts of the sample packaging (such as the upper, middle and lower parts) separately when sampling from the original sample for measurement.[7]

4.3.3 Measurement Condition

As the XRD method is relatively simple, this section describes the selection principle of optimal measurement conditions when measuring the crystal structure of graphene powder and related materials:

(1) **Tube voltage and tube current:** Tube voltage and tube current shall not exceed the maximum tube voltage and tube current

specified for the X-ray tube used; some instruments are expressed in terms of maximum power used.

(2) **Slit width:** The types of slits are divergent slit, anti-scattering slit, receiving slit and soller slit. In general, the slit width has an effect on the diffraction intensity and resolution. The larger the width, the greater the diffraction intensity, but the poorer the resolution. Conversely, the smaller the width, the smaller the diffraction intensity, but the better the resolution.

Select the appropriate slit width so that the X-rays can fully illuminate the sample measurement surface throughout the entire measurement process. The size of the divergence slit shall satisfy the conditions, that is, the width of the irradiated area on the sample surface obtained from Equation (4.7) is no larger than the width of the mounting window hole of the sample carrier. Anti-scattering slits generally use the same size as the divergence slits:

$$L = \alpha R / \sin \theta \qquad (4.7)$$

where L is the width of the irradiated area on the sample surface (mm), α is the divergence slit angle (°), R is the goniometer radius (mm) and θ is the Bragg angle (°).

(3) The spectral range is 5°–60°.
(4) The acquisition mode is continuous scanning or real-time acquisition.
(5) Spectral acquisition speed is (4°–8°)/min, and length of time is over 10 min.
(6) Spectral step width (for a scanning-type X-ray diffractometer) can usually be set at 0.02°, and a larger step width can be selected for graphene samples with a larger peak width. The spectral step width shall not be larger than one-third of the half-peak width of the most acute peak.

Collect the sample according to the above measurement parameters, and obtain the diffraction pattern of the sample. Repeat the sample measurement no less than three times.

4.3.4 Spectrum Analysis and Data Processing

In addition to obtaining the spectrum under the optimal measurement conditions, modern advanced measurement technology is equally important to the data processing of diffraction patterns. The data processing of XRD plots generally consists of the following three steps:

(1) **Smoothing processing:** Smooth each plot once with eleven points.
(2) **Background subtraction processing:** Background subtraction is required when the baseline is not level due to fluorescence peak, etc.
(3) **Peak searching:** Mark the data of diffraction peak angle 2θ, intensity, half-peak width, etc., calibrate the angle of the diffraction peak by the calibration curve obtained from the reference material, and calculate the interplanar spacing by Bragg's law.

The measurement result of each measurement sample shall include the average value and standard deviation of multiple measurements. In order to ensure the accuracy and reliability of measurement, it is necessary first to ensure that the equipment is within the calibration period.

4.4 Metrological Comparison

In order to ensure the operability and universality of the established measurement method as well as the comparability of measurement results, the established measurement methods need to undergo metrological comparison.[8] In the following, we introduce the domestic comparison of "The XRD method of measurement for graphene powder and related materials" led by NIM, to summarize the key factors of metrological comparison.

As the lead laboratory, the Materials Metrology Lab of China NIM selected the graphene powder samples A and B which met the homogeneity requirements as the comparison samples (see Fig. 4.9 for the XRD plot) and provided them to the comparison laboratories. The institute provided a comparison program based on scientific

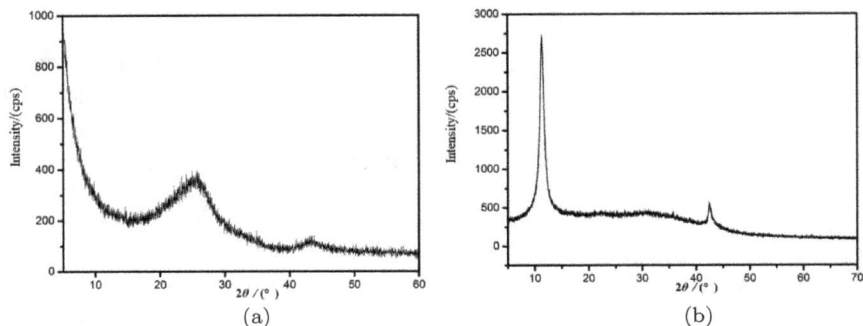

Fig. 4.9.   XRD figure of graphene powder materials: (a) comparison sample A; (b) comparison sample B.

research, which covers all aspects of equipment calibration methods, certified reference materials, measurement methods and data analysis (see Sections 4.2 and 4.3 for relevant information). It convened eight laboratories of the same level to carry out metrological comparison tests. The institute performed data analysis, determined reference value, evaluated the uncertainty in the collected results of the laboratories and finally gave the evaluation of the discreteness of measurement results.

The lead laboratory provided the silicon dioxide reference material SRM1878b released by NIST. Each unit measured the reference material before comparing it with the sample and then used the equipment after measuring the standard material to measure and compare sample A and the comparison sample B; each sample has three parallel samples (from the top, middle and bottom of the sample bottle, respectively).

Use the measurand value of certified reference materials in each unit as the vertical coordinate while the standard value of certified reference materials as the horizontal coordinate and perform linear fitting of the calibration curve. Take, for example, Unit 5 silicon dioxide reference material; see Table 4.2 for the measurement result. The linear fitting result is $y = 0.99995x - 0.00599$ and the linear correlation coefficient $r = 1.00000$.

According to the calibration curve obtained by instrument calibration, substitute y into the calibration curve as the measured value of the comparison sample and calculate the corresponding value as the calibration value of the measurement result. See Table 4.3 for

Table 4.2.   Measurement results of Unit 5 silicon dioxide reference material.

2θ standard value/(°)	2θ measurand value/(°)
20.860	20.856
26.639	26.631
36.543	36.534
42.449	42.441
59.958	59.947
90.828	90.811
106.589	106.590
120.119	120.102

Table 4.3.   Calibration values of comparison sample B for Unit 5 measurement.

Sample number, calibration value	Number of measurements	2θ measurand value/(°)	2θ calibration value/(°)
B1	1	11.929	11.935 6
	2	12.052	12.058 6
	3	12.154	12.160 6
B2	1	11.827	11.833 6
	2	11.991	11.997 6
	3	12.134	12.140 6
B3	1	11.847	11.853 6
	2	11.970	11.976 6
	3	12.113	12.119 6
Mean value		12.0019	12.008 5
Standard deviation		0.1207	0.1207

the obtained calibration values of comparison sample B for Unit 5 measurement.

When evaluating the dispersion of the measurement results from the various comparison laboratories, the normalized deviation value E_n needs to be calculated according to Equation (4.8):

$$E_n = \frac{Y_{ji} - Y_{ri}}{ku_i} \tag{4.8}$$

where k is the coverage factor, usually $k = 2$, and u_i is the standard uncertainty of $Y_{ji} - Y_{ri}$ on the ith measurement point.

It is clear from Equation (4.8) that the calculation of the value E_n is closely related to the uncertainty of measurement results in each laboratory E_n. The evaluation of uncertainty in measurement is described as follows.

Analyzing the entire measurement process, we can see that the uncertainty of measurement results mainly comes from the uncertainty introduced by calibration curve u_1, the uncertainty introduced by measurement repeatability u_2, the uncertainty introduced by the standard material u_3 and the uncertainty introduced by the sample homogeneity u_4:

(1) **Uncertainty introduced by calibration curve u_1:** See Equation (4.9) for uncertainty introduced by calibration curve:

$$u(x_0) = \frac{S}{b}\sqrt{\frac{1}{p} + \frac{1}{n} + \frac{(x_0 - \bar{x})^2}{\sum_{i=1}^{n}(x_i - \bar{x})^2}} \qquad (4.9)$$

where S is the residual standard deviation, $S = \sqrt{\frac{\sum_{i=1}^{n} y_i - (a+bx_i)^2}{n-2}}$; a is the intercept of the fitting line, b is the slope of the fitting line, p is the number of measurements of sample crystal plane (number of times), n is the total number of measurements of all facets of the reference material (number of times), x_0 is the mean value of a measurement sample (°), x is the mean value of the reference material (°), x_i is the standard value for the ith measurement of certified reference materials (°), $\bar{x}$ is the mean value of the standard values of the reference material (°) and y is the measurand value of certified reference material (°).

Calculate, for example, the measurement result of comparison sample B for Unit 5 measurement, $S = 0.0062$, $b = 0.99995$. Comparison sample B is measured for a total of 9 times, hence $p = 9$. Each of the 8 crystal planes of the reference material is measured once, hence $n = 8$. Calculate according to the equation, and the uncertainty $u(x_0)$ of the calibration result of the total average value ($x_0 = 12.0085$) after the calibration of the measured value of Sample B is 0.0044°.

(2) **Uncertainty u_2 introduced by measurement repeatability:** The uncertainty introduced by measurement repeatability can be expressed by calculating the standard deviation of repeated measurements. Calculate, for example, the measurement result of comparison sample B for Unit 5; the standard deviation of nine measurements is $u_2 = 0.1207°$.

(3) **Uncertainty u_3 introduced by certified reference materials:** In the reference material certificate of tested reference material NIST SRM 1878b, the value of a is (0.491 406 $\pm$ 0.000 020) nm, the value of c is (0.540 554 $\pm$ 0.000 020) nm and the uncertainty introduced by certified reference materials at $2\theta = 12.00085°$ is calculated to be 5.01×1^{-6}.

(4) **Uncertainty u_4 introduced by sample homogeneity:** Using the calibration values of measurement results of eight participating units as within-group values, the calibration values of measurement results between the participating units are used as the inter-group values, and the uncertainty introduced by the homogeneity of samples was calculated through F tests to evaluate the homogeneity of samples. The calculated results of homogeneity of comparison sample B is shown in Table 4.4. The uncertainty introduced by homogeneity is 0.665 7°.

combined uncertainty $u_c = \sqrt{u_1^2 + u_2^2 + u_3^2 + u_4^2}$; expanded uncertainty $U = ku_c$ $(k = 2)$.

In summary, take, for example, the measurement result of comparison sample B for Unit 5, the composite uncertainty is 0.6755°, and the extended uncertainty is 1.3531° $(k = 2)$.

Participating units provided evaluation of uncertainty and calculation of normalized deviation of data E_n, the result is shown in Tables 4.5 and 4.6 and Figures 4.10 and 4.11. It is clear from Tables 4.5 and 4.6 that the uncertainty levels of measurement results in different laboratories are similar, so the arithmetic mean results from eight laboratories were used as reference values, and E_n value statistical method was used to evaluate the dispersion of measurement results. According to JJF1117-2010 *Metrological comparison*, if $|E_n| \leq 1$, the difference between the reference laboratory measurement and the reference value was within reasonable expectation, and

Table 4.4.   Calculated results of homogeneity of comparison sample B [unit: $(°)$].

m					n					X
	1	2	3	4	5	6	7	8	9	
1	11.9814	12.1779	11.9302	11.9872	12.0712	12.0504	11.9003	12.0011	12.0894	12.0210
2	12.5836	12.6494	12.7042	12.5676	12.6394	12.6992	12.6195	12.6783	12.7241	12.6517
3	11.3456	11.3426	11.3146	11.5368	11.5358	11.5398	11.6966	11.7046	11.5406	11.5063
4	11.0049	11.0059	10.9179	11.0099	11.1239	11.1089	11.0949	11.0259	10.9089	11.0223
5	11.9356	12.0586	12.1606	11.8336	11.9976	12.1406	11.8536	11.9766	12.1196	12.0085
6	11.0277	11.0287	10.9407	11.0327	11.1467	11.1317	11.1177	11.0487	10.9317	11.0451
7	10.3995	10.4709	10.5275	10.8921	10.8155	10.9912	10.4022	10.4492	10.3207	10.5854
8	11.5878	11.6618	11.6308	11.6428	11.5828	11.5898	11.4788	11.5958	11.6168	11.5986

Total mean value $\bar{x}$			11.5549					Sum of squares between groups Q_1		28.0226
Sum of squares within a group Q_2			0.9702					Number of measurements N	72	
Degree of freedom between groups ν_1			7					Degree of freedom between groups ν_2	64	
S_1^2			4.0032					S_2^2		0.0152

Table 4.5.   Measurement results of comparison sample A.

Unit no. of reference laboratories	$2\theta/(°)$	Extended uncertainty $(k = 2)/(°)$	E_n
1	26.1615	0.8867	0.05
2	26.7873	0.7507	0.83
3	25.9088	0.7594	−0.25
4	26.3797	0.7005	0.35
5	26.0624	0.7393	−0.06
6	25.6665	1.0438	−0.41
7	25.7759	1.0649	−0.30
8	26.1730	1.0886	0.05
Reference value	26.1144	0.3155	—

Table 4.6.   Measurement results of comparison sample B.

Unit no. of reference laboratories	$2\theta/(°)$	Extended uncertainty $(k = 2)/(°)$	E_n
1	12.0210	1.3425	0.33
2	12.6517	1.3769	0.75
3	11.5063	1.3631	−0.03
4	11.0223	1.3403	−0.37
5	12.0085	1.3531	0.32
6	11.0451	1.3403	−0.36
7	10.5854	1.4198	−0.65
8	11.5986	1.3355	0.03
Reference value	11.5549	0.4805	—

the comparison results were acceptable. It can be seen from Tables 4.5 and 4.6 and Figures 4.10 and 4.11 that the results from the reference laboratories were all acceptable.

The consistency of measurement results in different laboratories was obtained through metrological comparison, which verified the reliability, operability and universality of the established measurement method and showed the effectiveness and reliability of the established measurement method.

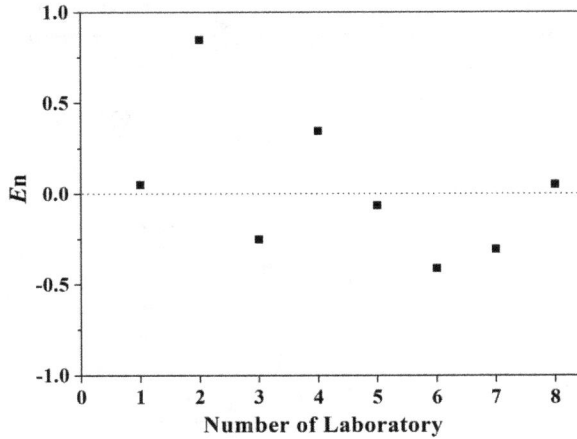

Fig. 4.10. E_n value of measurement results of comparison sample A.

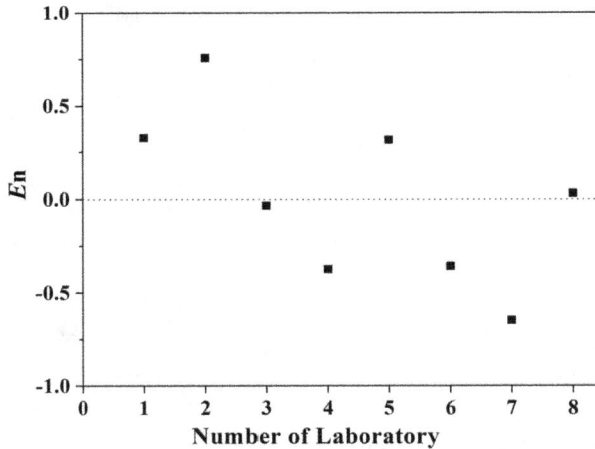

Fig. 4.11. E_n value of measurement results of comparison sample B.

4.5   Standard Method

Through the studies from equipment calibration to sample preparation, sampling and measurement conditions, and the equivalent consistency of measurement results verified by measurement comparison described in Sections 4.2–4.4, the reliability, operability and applicability of the measurement method are verified, which is suitable for scientific research and industrial promotion and application. Therefore, in accordance with the rules provided by GB/T 1.1-2009 *Part I: Guidelines for standardization: Structure and preparation of standards* and T/CSTM00002-2019 *General Rules for the compilation of standards of measurement methods*, Chinese Society for Testing and Materials (CSTM), Zhongguancun, drafted and finally issued the group standard T/CSTM 00166.2-2019 *Characterization of graphene materials, Part II: X-ray diffraction method*. This standard provides a unified measurement basis for measuring and testing the crystal structure of graphene materials.

As one of a series of standards for characterizing the crystal structure of graphene-related materials, this standard method will provide technical guidance for the production and research of graphene materials, in combination with other standards of this series, such as Raman spectroscopy, atomic force microscopy and transmission electron microscope.

4.6 Summary

Two-dimensional graphene-related materials are the collective name for graphene and its derivatives with layers less than 10; its crystal structure varies with the derivative structure. XRD is one of the important technologies for measuring the crystal structure of graphene-related two-dimensional materials; the accuracy of XRD measurement result is critical to the judgment of its structure. Equipment calibration can guarantee the accuracy and reliability of the equipment, while comparison of measurements can guarantee the consistency and operability of measurement methods. Based on the verification of the reliability of the measurement method, the measurement method can be upgraded to the standard method, to provide scientific research and production units a unified basis for measurement and accurate measurement data, which helps shorten the research and development time and improve product quality.

This chapter has introduced the measurement principle and method of XRD technique. Given the high X-ray energy, the penetration depth of the X-ray ranges from a few microns to tens of microns. But for sub-nano to nanothickness graphene film samples, the X-ray passes through the film material and penetrates into the substrate, so the diffraction signal from the conventional symmetric diffraction is the superimposition of base diffraction signal and thin film material diffraction information, and the intensity of the diffraction signal is directly related to the diffraction volume. Therefore, most of the information in the spectrum will come from the substrate material of the film sample, while the thin film material itself accounts for only a small part of the information. Hence, the grazing incidence mode (the incident angle is very low and remains constant during the measurement) is needed. That allows the X-rays to travel very long distances in thin film materials but at very low actual longitudinal depths, thus realizing the goal of increasing the proportion of the information of the thin film material in the spectrum or even covering the spectrum entirely with thin-film material information. Therefore, for graphene powder and graphene film, we need to select different types of XRD. When the film thickness is measured by grazing incidence, equipment calibration and measurement methods need to be re-established. The equipment calibration of grazing incidence method is listed in JJF 1613–2017.[9]

References

[1]   Ma L. *Modern X-ray Polycrystal Diffraction: Experimental Techniques and Data Analysis.* Beijing: Chemical Industry Press, 2004.

[2]   Huh S H. Thermal reduction of graphene oxide. In S Mikhailov (eds.), *Physics and Applications of Graphene — Experiments* (pp. 73–90). InTech, 2011.

[3]   Dreyer D R, Murali S, Zhu Y W, *et al.* Reduction of graphite oxide using alcohols. *Journal of Materials Chemistry*, 2010, 21(10): 3443–3447.

[4]   Ren L, Gao H. X-ray tracing of X-ray diffractometer. *Metrological Technology*, 2012, (8): 3–5.

[5]   Ren L, Cui J. Angle tracing of the X-ray diffractometer. *Metrological Technology*, 2012, (3): 48–51.

[6]   T/CSTM 00166.2–2019 Characterization for graphene materials Part 2 X-ray diffraction.

[7]   JJG 1343–2022 Characterization, Homogeneity and Stability Assessment of Reference Materials.

[8]   JJF 1117–2010 Measurement Comparison.

[9]   JJF 1613–2017 Calibration Specification for Thin Film Thickness Measurement Instruments by Grazing Incidence X-Ray Reflectivity.

https://doi.org/10.1142/9789811295713_0005

Chapter 5

Measurement of Thickness of Graphene-Related Materials by Atomic Force Microscopy (AFM)

Tianjia Bu

5.1 Overview

The excellent properties of graphene are attributed to its unique ultra-thin two-dimensional structure. Therefore, an accurate and reliable characterization of the structural characteristic parameters of graphene is the key basic element to ensure the R&D of graphene and even its industrialization. In 2004, Andre K. Geim *et al.* separated and discovered graphene for the first time. As one of the measurement methods, atomic force microscope (AFM) technology proved the two-dimensional single-layer structure of graphene. Among the structural characteristic parameters of graphene, the thickness of an atom scale is by far the most intuitive and the most important parameter to determine its intrinsic properties. For this very reason, among the graphene national quality infrastructure (NQI) technology survey results, accurate and consistent method for measuring the thickness of graphene-related materials has been evaluated as one of the most urgently needed metrology technologies.[1] AFM[2] technology is the most widely used technology at present for characterizing three-dimensional surface topography. With a lateral resolution of 0.1 nm and a longitudinal resolution of 0.01 nm, it is considered the most powerful, most direct and effective method to characterize the topography and thickness of graphene. However, because of the super large specific surface area of graphene-related materials, the interactions with the substrate and the tip effect of AFM, etc., the thickness of single-layer graphene reported in the literature is around (0.4–1.7) nm (Table 5.1), which is sharply different from the theoretical thickness of single-layer graphene (0.34 nm). Therefore, in order to determine whether the tested sample is graphene with few layers or graphite, there is an urgent need to develop AFM technology to accurately measure the thickness of graphene-related materials so as to meet the needs of scientific research and industry. This chapter mainly introduces the calibration and traceability of AFM instrument and measurement methods involved in the accurate measurement of the thickness of graphene-related materials, as well as how to achieve equivalence and consistency of measured value through international comparison.

5.2 Principle of Atomic Force Microscopy Technology

AFM is one of the most widely used surface observation and research tools in the family of scanning probe microscopes (SPMs), its basic composition is shown in Fig. 5.1.

AFM technology can not only be imaged at the atomic scale, but it also has a simple sample preparation process and a variety of test environments including vacuum, atmosphere and solution. Besides, it is able to study a wide range of subjects, including conductors, semiconductors, insulators and other materials with different conductivity, inorganic, organic polymer and other chemical materials, cells, DNA and other biological samples. Therefore, although AFM has been available for a short time, its application theories and technologies have developed rapidly. As one of the important research techniques that promote the progress of material science, electronic technology, life science and surface science, it has a broad prospect of development.

The imaging principle of AFM[2] is to examine the surface morphology of the sample by using an elastic microcantilever with a probe. When the tip on the probe is scanned on the surface of the sample, the interaction force between the tip and the sample will be sensed on the microcantilever, inducing microcantilever deformation, which is detected by the laser reflection system to obtain the surface morphology of the sample. For different types of samples, the AFM

Table 5.1.   Summary of literature on graphene-related materials' thickness measurements using AFM technology.

Thickness/nm	AFM scanning mode	Types of graphene materials	Layers	Substrates	References
0.4–0.9	Tapping mode	Mechanical stripping method for preparation of graphene	1	Mica	[3,4]
0.4–1.7	Tapping mode	Mechanical stripping method for preparation of graphene	1	Si/SiO_2	[5]
1.8	Tapping mode	Graphene grown by Chemical Vapor Deposition (CVD) method	1	Si/SiO_2	[6]
1.44	Tapping mode	Reduced graphene oxide	1	HOPG	[7]
0.8–1.5	Tapping mode	Reduced graphene oxide	1	Si/SiO_2 (300 nm)	[8]
1.1 ± 0.1	Tapping mode	Graphene oxide and reduced graphene oxide	1	HOPG	[9]
0.8–1.1	Tapping mode	Graphene oxide	1	Si/SiO_2 (300 nm)	[10]
0.9–1.7	Tapping mode and contact mode	Graphene oxide and reduced graphene oxide	1	HOPG	[11]
0.9 ± 0.2	Tapping mode	Mechanical stripping method for preparation of graphene	1	Si/SiO_2	[12]
1.19 ± 0.1	Tapping mode	Mechanical stripping method for preparation of graphene	1	Si/SiO_2 (300 nm)	[13]
0.9	Contact mode	Mechanical stripping method for preparation of graphene	1	Si/SiO_2	[14]
0.4–1	Contact mode	Mechanical stripping method for preparation of graphene	1	Si/SiO_2	[15]
0.7	Contact mode	Mechanical stripping method for preparation of graphene	1	Si/SiO_2 (300 nm)	[16]
1	Contact mode	Mechanical stripping method for preparation of graphene	1	$Si\{111\}$	[17]
0.4	Ultra-high vacuum mode	Mechanical stripping method for preparation of graphene	1	Si/SiO_2 (300 nm)	[18]

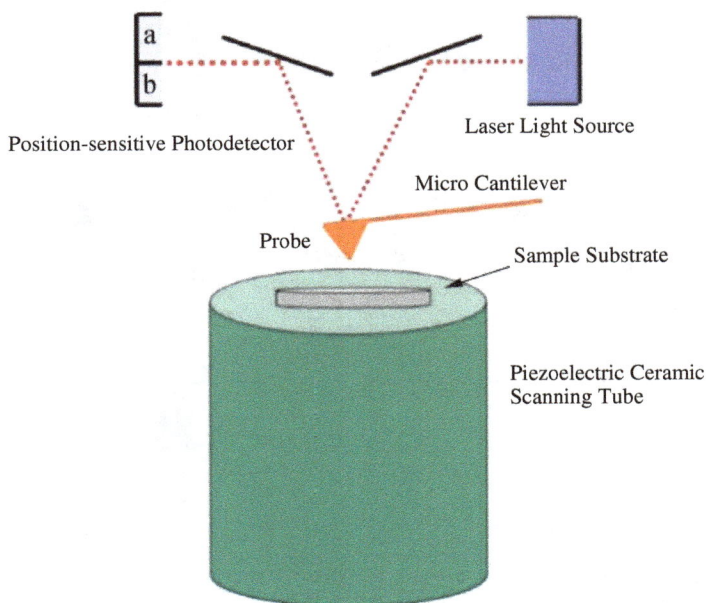

Fig. 5.1.　Basic composition of atomic force microscopy.

imaging modes are divided into contact mode, tapping (intermittent contact) mode and non-contact mode. Considering the flexibility of graphene-related materials and the resolution requirements, the tapping (intermittent contact) mode is mostly used for measurement.

5.3　Calibration and Traceability of AFM Equipment

AFM is an instrument with sub-nano spatial resolution that can detect the physical characteristics of the measured surface and the near-surface area at the atomic level. The metrology institutes of various countries actively carry out traceability research to ensure the accuracy and comparability of AFM measurement results. With the support of the National Natural Science Foundation of China, National Institute of Metrology (NIM, China) established a metrological AFM in collaboration with Physikalisch-Technische Bundesanstalt (PTB),[19] tracing the geometric quantity to the wavelength of light (SI unit of length). This has provided a scientific basis for the absolute measurement and metrological calibration of nanometer value. The measurement range of the main technical indexes of the metrological standard equipment is 70 μm in x direction and 15 μm in y and z directions, and the measurement resolution is about 1.2 nm in x direction and 0.25 nm in y and z directions. In order to carry out the calibration and absolute measurements on the three axes of the metrological AFM, an integrated laser interferometric three-dimensional measurement system is installed on it. The resolution of the micro-fiber conduction laser interferometric 3D measurement system is 0.1 nm, which can simultaneously measure the relative shift of three (x, y and z) moving directions in AFM between the measurement component and the stationary part of the scanning component, and the measurement uncertainty is (2–3) nm. The interference system also has a pulse output of $\lambda/2$ with a measurement uncertainty of $U_{95} = 1$ nm. When calibrating the AFM, this pulse output is used. The reading number and $\lambda/2$ pulse of the interference system can be input into the computer simultaneously. Detailed research on traceability technology is shown in references.[20]

In order to pass the value of AFM measurement standard device that is traceable to SI units to the end user, it is necessary to calibrate the end user equipment in accordance with the calibration specification (*Calibration Specifications of Scanning Probe Microscopy*) (JJF1351-2012), which can ensure the unity of value. According to the specification JJF1351-201, JJF 1351-2012 the end user's AFM equipment is calibrated using the physical carrier of the standard instruments with line interval, line width and step height that carry the accurate value of AFM metrological standard device. Five metrological characteristics should be calibrated: z-direction drift, measurement error of y-axial displacement, measurement error of z-axial displacement, measurement repeatability, as well as x and y coordinate orthogonality error. Different calibration items of scanning probe microscopy use different standards, as listed in Table 5.2.[21] The topography and three-dimensional drawing of the nanostep template are shown in Fig. 5.2,[21] and one-dimensional and two-dimensional nanowire spacer templates are shown in Fig. 5.3.[21] The measurement uncertainties introduced in the calibration process are evaluated by such factors as the measurement standard equipment, the quality of the surface structure of the measured object, the thermal expansion coefficient of the material and the degree of contamination.

Table 5.2.   Scanning probe microscopy calibration items and corresponding standard equipment.[21]

No.	Calibration item	Standard equipment and technical requirements	
		Standard equipment	Technical requirements
1	Scanning probe microscopy z-direction drift	Nanostep template	$U = 4\,\mathrm{mm} + 5 \times 10^{-5}h$, $k = 2$ h is the height of steps
2	Measurement error of x- and y-axial displacement	Nanowire spacer template	Maximum allowable error: $\pm 1\,\mathrm{nm}$
3	Measurement error of z-axial displacement	Nanostep template	$U = 4\,\mathrm{mm} + 5 \times 10^{-5}h$, $k = 2$ h is the height of steps
4	Repeatability of scanning probe microscopy measurements	Nanostep template	$U = 4\,\mathrm{mm} + 5 \times 10^{-5}h$, $k = 2$ h is the height of steps
5	x and y coordinate orthogonality error	Two-dimensional nanowire spacer template	Maximum allowable error: $\pm 0.1°$

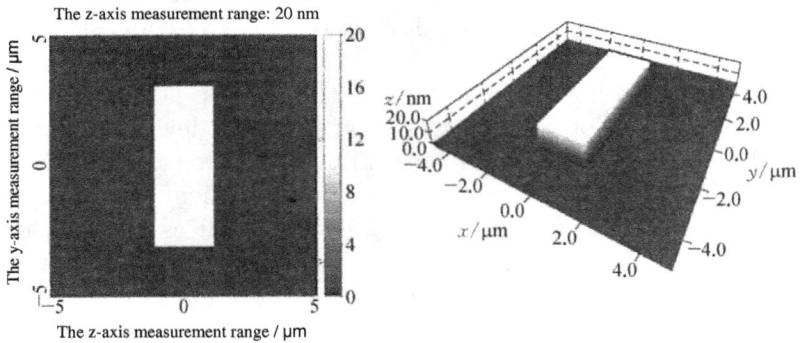

Fig. 5.2.   Topography and three-dimensional drawing of nanostep template.[21]

It should be noted that during the calibration process, it is necessary to ensure that the calibrated quantity value range meets the quantity value range of the measured object. At present, the smallest scale of nanostep certified reference material (CRM) in China is

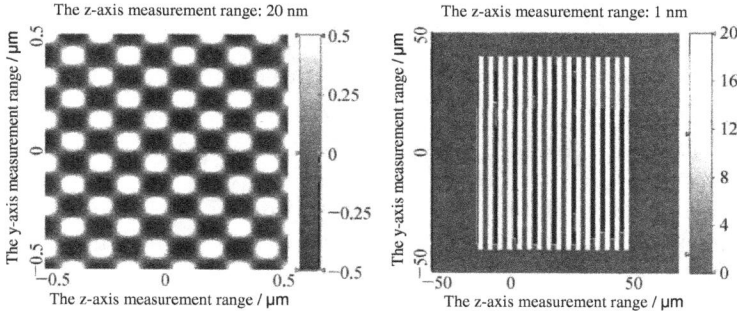

Fig. 5.3.   Topography of one-dimensional and two-dimensional nanowire spacer templates.[21]

a single-step template with a nominal height of 5 nm prepared by atomic layer deposition (ALD) technology, in other words, the minimum calibration range of z-axis that can meet is 5 nm, which cannot meet the measurement requirements of graphene-related materials' thickness (0.334–3.4 nm). Therefore, in the research of thickness measurement technology of graphene-related materials, it is necessary to further reduce the lower limit of the AFM value calibration range to less than 1 nm scale range. Advanced material metrology laboratory of NIM (China) proposed calibration methods of scales smaller than 1 nm through calibration curves method established by CRMs in different ranges of value: (1) Select CRMs of three heights, namely the quartz step height standards of VLSI Standards Inc, strontium titanate and silicon {111} monatomic steps, corresponding respectively to the heights of 8.9 nm, 0.39 nm and 0.31 nm, (2) calibrate the three CRMs respectively with AFM, and (3) establish calibration curves to calibrate AFM. The calibration curve is used to calibrate the values below 10 nm; the evaluation equation of measurement uncertainty introduced by calibration curve is as follows:

$$u(x_0) = \frac{S}{b} \sqrt{\frac{1}{p} + \frac{1}{n} + \frac{(x_0 - \bar{x})^2}{\epsilon_{i=1}^n (x_i - \bar{x})^2}} \qquad (5.1)$$

among

$$S = \sqrt{\frac{\epsilon_{i=1}^n [y_i - (a + bx_i)]^2}{n - 2}}, \text{ nm}; \qquad (5.2)$$

where a is the intercept of a fitting line (nm), b is the slope of a fitting line (dimensionless), S is the residual standard deviation (nm), p is the number of measurements of the sample to be tested (number of measurements), n is the number of measurements of CRM (number of measurements), x_0 is the measured mean value of test sample (nm), x is the measured mean value of CRM (nm), x_i is the standard value of the ith measurement of CRM (nm), $\bar{x}$ is the mean value of the standard values of CRM (nm) and y is the measured value of CRM (nm).

5.4   Establishment of Measurement Method

In order to obtain equivalent and consistent measurement results, it is necessary to ensure the operability and universality of the measurement method on the basis of AFM equipment calibration. As shown in Table 5.1, the interactions between AFM substrates and graphene-related materials, AFM probe and graphene-related materials, as well as AFM data analysis and processing methods and other factors affecting the measurement results are reported in the literature. Therefore, this section introduces the influence of AFM technology on various parameters and the selection of methods for measuring the thickness of graphene-related materials.

5.4.1   Selection of AFM Scanning Mode

Tapping (intermittent contact) mode is commonly used to characterize graphene-related materials by AFM technology. This mode is between contact mode and non-contact mode, including two principles: amplitude modulation and external force modulation. In the process of measurement, AFM measurement uses cantilevers with probes which oscillate at a fixed frequency above the surface of graphene-related materials and the tip briefly touches or strikes the surface of the sample periodically. Researches show that the optimized tapping (intermittent contact) mode can effectively eliminate the interaction between the nanomaterials and the tip,[22–24] to ensure the reliability of the scanned image data. Some researchers have also tried to characterize the thickness of graphene films with contact mode to determine the number of layers, but differences in height were observed between the forward and reverse scan results.[5] These differences are attributed to the higher lateral forces between the material and the tip. Therefore, tapping (intermittent contact) mode is recommended when AFM technology is used to measure the thickness of graphene-related materials.

5.4.2   Effect of Substrates for AFM Measurement

The actual thickness of graphene-related materials characterized by AFM technology is often much larger than the theoretical thickness (0.34 nm) of the graphite monolayer. This is due to the existence of adsorbent on the surface of graphene, the interaction between different substrates and graphene materials has an impact on the thickness. Graphene forms ripples in suspension or on most supports, resulting from maintaining thermodynamic stability on its independence. However, Lui[25] and other researchers have found that when graphene was deposited on a mica surface under certain conditions, the change in apparent height of graphene less than 25 pm was observed through high-resolution AFM (Fig. 5.4), indicating that the intrinsic ripples of graphene for its own stability were suppressed, displaying the state of superflat graphene. We have also found and demonstrated that during our experiments. Therefore, a newly dissociated mica as a substrate is recommended when AFM technology is used to measure the thickness of graphene-related materials.

5.4.3   Effect of AFM Measurement Parameters

The setting of AFM measurement parameters is particularly important to the result of material thickness measurement, which determines the interaction between AFM measuring probe and graphene-related materials as well as the feedback accuracy of image scanning, thus affecting the true accuracy of the scanned image. Shearer[23] and other researchers have used commercial probes and carbon nanotube-modified AFM probes combined with PeakForce tapping mode to study the thickness of graphene film. It was found that an adsorbed layer between graphene and the substrate and the imaging force, particularly the amount of pressure applied by the tip

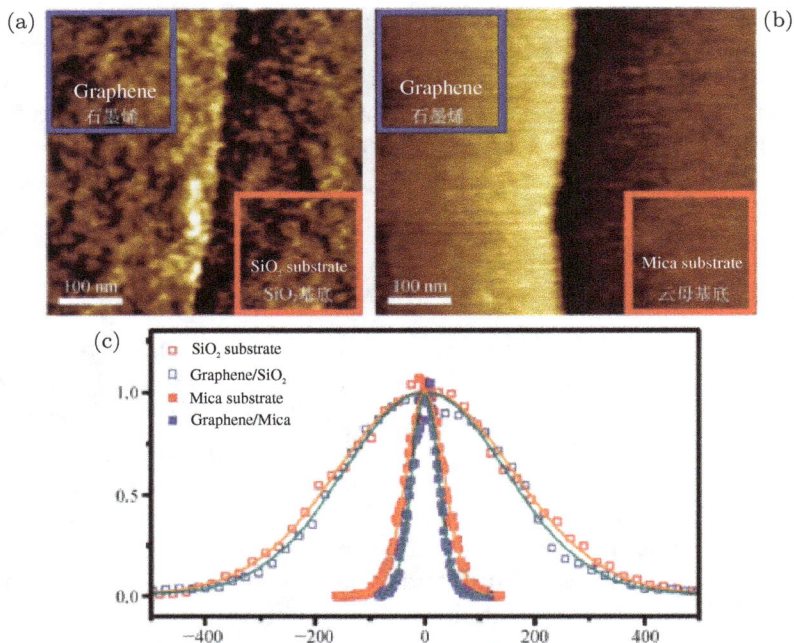

Fig. 5.4.   AFM height maps of graphene on different substrates and correspond-ing height histograms[25]: (a) the AFM height map of graphene on the SO_2 substrate; (b) the AFM height map of graphene on the mica substrate; (c) the corresponding height histogram (from inside to outside) of graphene on the mica substrate, blank mica substrate, graphene on SiO_2 substrate and blank SiO_2 substrate.

of the cantilever, were the key factors in accurately measuring the thickness of graphene (Fig. 5.5). The measurement error of the dif-ference between the height of graphene and the substrate could be effectively reduced from (0.1–1.3) nm to (0.1–0.3) nm.

In addition, researchers also found that in amplitude modula-tion mode, the amplitude setting value and the driving amplitude played a crucial role in the accuracy of the thickness results of graphene-related materials, and too large or too small amplitude parameters could lead to a thickness deviation of up to 2 nm. By studying the interaction between the probe and graphene-related materials, image scanning feedback accuracy and other parameters, the amplitude setting value for the probes and equipment used is set at (100–120) mV, and the drive amplitude is set at (80–140) mV.[26]

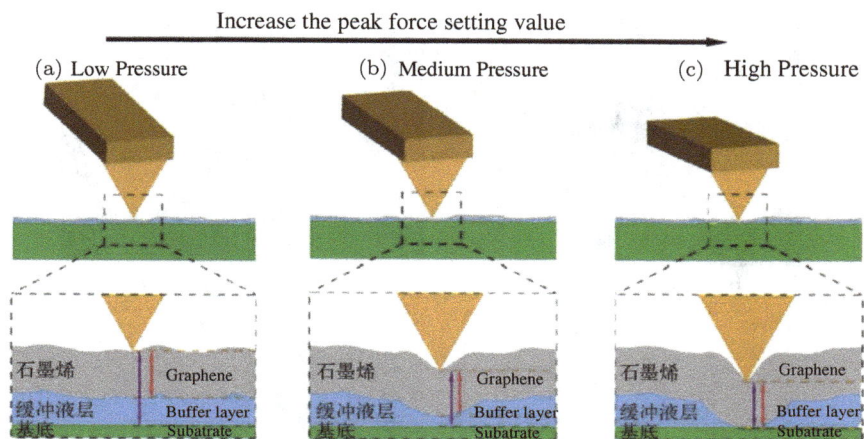

Fig. 5.5.   Schematic diagram of the mechanism of the peak force setpoint and AFM imaging accuracy.[23] When the applied pressure rises from low (a) to medium (b) and to high (c), the pressure on AFM tip can destroy the bottom buffer layer to measure more accurately the height of graphene.

5.4.4   Methods of AFM Data Analysis and Processing

For graphene-related materials with a thickness of less than 1 nm, any deviation introduced by factors such as pollutants or external noise will have a great impact on the accuracy of thickness measurement. Usually, the choice of the center line of the upper and lower steps is extremely important for the step height difference method in AFM data processing which is used to calculate the film thickness. However, as can be seen from Fig. 5.6(d), the value of the noise signal of AA and AB steps is greater than the value of the step height itself (the noise of AB step is less but still reaches 2.6 nm, greater than the step height 1.52 nm). Therefore, how to accurately determine the step height value corresponding to the substrate and sample scanning data is very critical. Yao and other researchers from Advanced material metrology laboratory of NIM (China),[26] analyzed the data of AFM technological measurement of step baseline in the range of 1 nm value via probability distribution statistics method, and measured the thickness and number of layers of the multi-layer graphene films grown by CVD method on silicon substrate. The experiment used a probability distribution statistical method to determine the

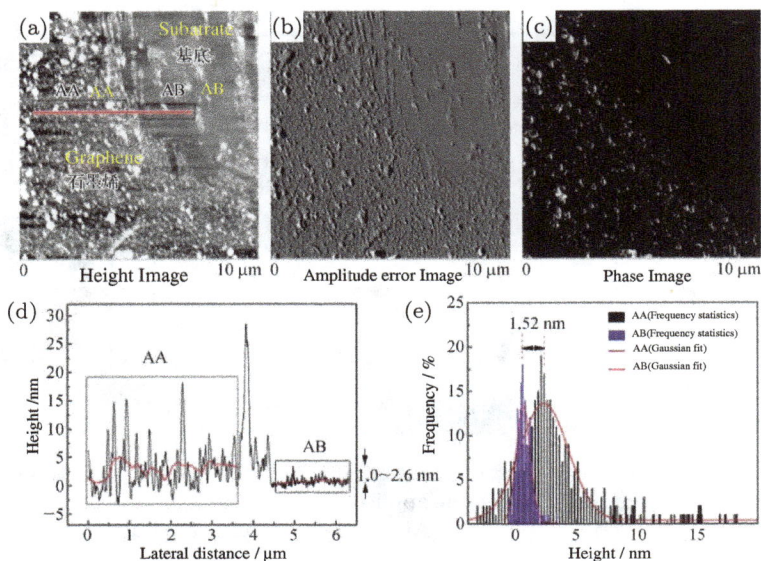

Fig. 5.6. AFM measurement results of single-layer (1L) graphene film[26]: (a) morphology diagram; (b) amplitude error diagram; (c) phase diagram; (d) height profile corresponding to the red line in (a); (e) histogram of height difference between graphene (AA region) and substrate (AB area).

height difference between graphene-related materials and the substrate, the calculated thickness between the single-layer graphene film (1L) and the substrate is (1.51 ± 0.16) nm, much larger than the theoretical thickness of the single-layer graphene (Fig. 5.6). This could be caused by a van der Waals force between the graphene film and the substrate combined with contamination in the transfer process. In the same vein, they studied the thickness of multi-layer graphene films grown directly by CVD method and analyzed the thickness between the two-layer (2L) and four-layer (4L) graphene films and the substrate, the results are respectively (1.92 ± 0.13) nm (Fig. 5.7) and (2.73 ± 0.10) nm (Fig. 5.8).

Considering that the transfer process did not introduce contamination between the graphene film layers, they proposed the subtraction method to calculate the single-layer (1L) thickness of four-layer (4L) graphene films: thickness (1L) = [thickness (4L) − thickness (1L)]/(4 − 1). The single-layer thickness of the four-layer graphene film is calculated to be $[(2.73 - 1.51)/3 \pm (0.16 + 0.10)/3]$ nm =

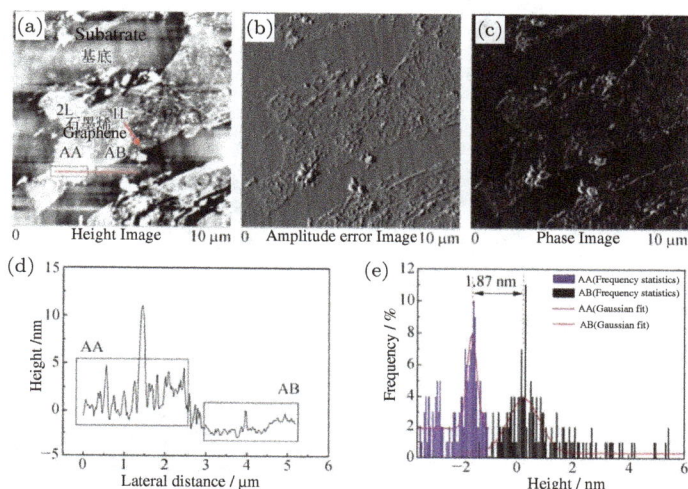

Fig. 5.7.   AFM measurement results of two-layers (2L) graphene film[26]: (a) morphology diagram; (b) amplitude error diagram; (c) phase diagram; (d) height profile corresponding to the red line in (a); (e) histogram of height difference between graphene (AA region) and substrate (AB area).

Table 5.3.   Different image smoothing processing parameters versus the corresponding single-layer graphene thickness results.[26]

Image smoothing processing parameter	Height difference/nm
0 order	4.01
1st order	1.63
2nd order	1.52
3rd order	−0.32

(0.41 ± 0.09) nm, basically consistent with the theoretical thickness value of the single-layer graphene (0.334 nm). It thus proves that the statistical method of probability distribution can eliminate the random error caused by human factors in establishing the baseline of the step and can effectively analyze the step baseline data of the 1 nm scale range of graphene materials to obtain the accurate film thickness.

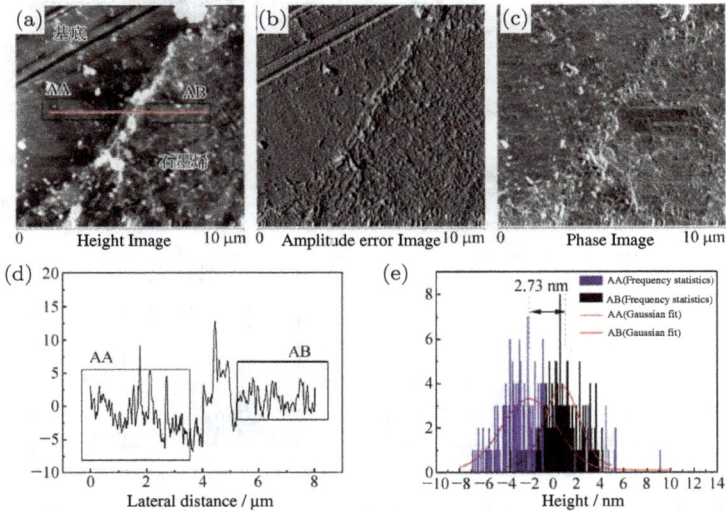

Fig. 5.8.   AFM measurement results of four-layer (4L) graphene film[26]: (a) morphology diagram; (b) amplitude error diagram; (c) phase diagram; (d) height profile corresponding to the red line in (a); (e) histogram of height difference between graphene (AA region) and substrate (AB area).

At the same time, the influence of the image smoothing processing parameters on the accuracy of the measurement results is proposed, showing that excessive image correction can lead to serious deviation in the results (Table 5.3). Therefore, it is suggested that the statistical method of probability distribution should be used for the data processing method for measuring the thickness of graphene materials by AFM technology, and the image smoothing parameters are first or second order.

5.4.5   Evaluation of Uncertainty

The uncertainty of measurement results mainly includes uncertainty of class A, uncertainty of class B and the uncertainty introduced by sample homogeneity.

(1) **Type A Uncertainty:** Type A uncertainty includes the uncertainty introduced by the test method, such as the uncertainty u_1 introduced by the measurement method. The uncertainty u_1 introduced by the measurement method (repeatability) is the

repeated measurement of the same sample several times, the standard deviation of the measurement results.

(2) **Type B uncertainty:** Type B uncertainty is mainly the uncertainty introduced by instrument calibration, which mainly includes the uncertainty u_2 introduced by CRMs (calibration sample) and the uncertainty u_3 introduced by the calibration curve in Section 4.3.2.

(3) **Uncertainty introduced by sample homogeneity:** Uncertainty introduced by sample homogeneity u_4 is measured at different positions of the sample according to the sampling principle and the standard deviation of the measurement results.

Since each uncertainty component is not correlated, the combined uncertainty of measurement results u_c is obtained by combining the square and root forms, that is,

$$u_c = \sqrt{u_1^2 + u_2^2 + u_3^2 + u_4^2} \tag{5.3}$$

For a normal distribution, when the confidence level is 95% and corresponding $k = 2$, the expanded uncertainty U is

$$U = ku_c = 2 \times u_c \tag{5.4}$$

5.5 Metrological Comparison

Metrological comparison is a metrological term with strict procedural requirements. The comparison can be used to validate the universality and operability of a standard measurement method when it is established, to check and evaluate the measurement ability of the laboratory after the establishment of the method and to verify and evaluate equivalent and consistent value of metrological standards (standard equipment or CRM). Advanced material metrology laboratory of NIM (China) led the international and domestic comparisons of AFM measurements of graphene oxide (GO) thickness, the comparison sample and the comparison scheme were consistent. The purpose of the domestic comparison is to verify the universality and operability of the established method in Section 5.4, while the purpose of the international comparison is to evaluate the equivalence and consistency of GO thickness measurements on mica substrates.

Fig. 5.9.   The packaged comparison sample.

5.5.1   Selection of Comparison Samples

According to *Metrological comparison*, the samples in the leading laboratory were first compared with the stability, uniformity, impact quantity and transportation characteristics according to the relevant requirements of measurement comparison.

Due to the variety of graphene-related materials, and the fact that there is no thickness-related CRM at present, comparison sample should be selected from a large number of samples to meet the comparison requirement as the reference sample. The comparison leading laboratory measured the parameter of thickness of different substrates including Si/SiO_2 substrate, Si substrate and mica substrate on various kinds of graphene-related materials produced by domestic and foreign graphene production enterprises, including a multi-layer graphene film on a Si/SiO_2 substrate, reduced GO powder material, reduced GO slurry material, GO powder material and GO slurry material, and determined through a homogeneity test the sample of GO spread on the newly dissociated mica substrate as the comparison sample.

The thickness of the comparison sample is only about 1 nm; any improper operation may affect the accuracy of the value. Therefore, the lead laboratory carried out research on the storage and packaging of the selected comparison samples and finally determined to use vacuum packaging and room temperature storage so that the comparison samples would not be damaged and contaminated when they were mailed to domestic and foreign laboratories (Fig. 5.9). During the above study, the comparison leading laboratory jointly carried out a bilateral comparison with the National Physical Laboratory (NPL) of UK, in order to determine the homogeneity and stability of the comparison samples finally sent. The results showed equivalence and consistency, which further determined the reliability of the comparison samples.

5.5.2   Domestic Comparison

To ensure the operability, universality of established measurement methods and the comparability of measurement results, a metrological comparison of the established measurement methods is required. This section introduces the domestic comparison of "Measurement of the thickness of GO flakes by AFM" led by Advanced material metrology laboratory of NIM (China) and highlights the key factors of metrological comparison.

As the comparison leading laboratory, Advanced material metrology laboratory of NIM (China) (1) screened out the comparison samples which meet the requirements of homogeneity and stability (Fig. 5.9) and provided them to the participating laboratories, (2) provided a comparison measurement instruction based on scientific research, including all aspects of instrument calibration methods CRMs, measurement methods and data analysis (Sections 5.3 and 5.4), (3) convened eleven laboratories with the same level to carry out the metrological comparison test, (4) analyzed data obtained from participating laboratory, selected reference value and evaluate uncertainties, and finally provided an evaluation of the discreteness of the measurement results. The comparison measurement instruction and test samples are uniformly distributed by comparison leading laboratory, adopting a star-shaped transfer mode (Fig. 5.10, refer to JJF 1117-2010 for specific methods and requirements).

Participants send the measurement data back to the leading laboratory as required. The data were compared and processed by the leading laboratory, including inspection of comparison data, data correction and statistics, reference value determination, comparison results from participants processing and evaluation, confirmation of

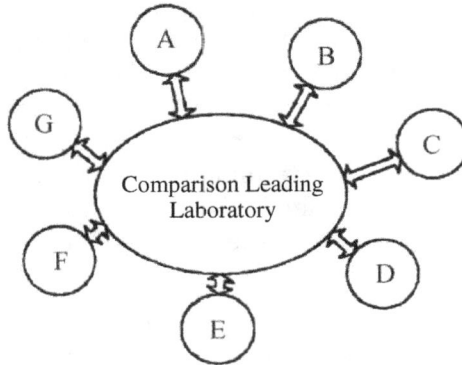

Fig. 5.10.   Schematic diagram of the star-shaped comparison mode.

abnormal or suspicious results, data preservation, data confidentiality, and drafting and revision of comparative summary report. The evaluation results of participants are shown in Fig. 5.11. Take the measurement results of NIM (China) as the central value and evaluate the discreteness of measurement results by E_n-value statistical method. According to ISO 13528-2009 *Statistical methods for capability verification using interlaboratory comparisons* and JJF1117-2010 *Metrological comparison*, the consistency of the measurement results and uncertainty of one participant is evaluated by normalized deviation E_n:

$$E_n = \frac{Y_{ji} - Y_{ri}}{k u_i} \qquad (5.5)$$

where k is the coverage factor, in general $k = 2$, and u_i is the standard uncertainty of $Y_{ji} - Y_{ri}$ at the ith measurement point.

When u_{ri}, u_{ji} and u_{ei} are unrelated or weakly related,

$$u_i = \sqrt{u_{ji}^2 + u_{ri}^2 + u_{ei}^2} \qquad (5.6)$$

where u_{ri} is the standard uncertainty of the reference value at the ith measurement point, u_{ji} is the standard uncertainty of measurement result at the ith measurement point of the jth laboratory and u_{ei} is the impact of the instability of transfer standard at the ith measurement point during the comparison of measurement results.

Principles for evaluating the consistency of comparison results are as follows:

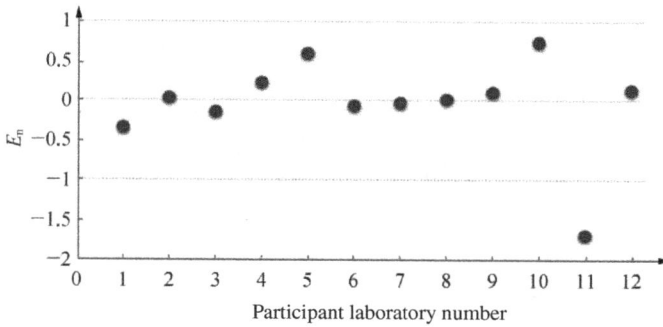

Fig. 5.11. Evaluation results of the reference units.

Table 5.4. Expanded uncertainty of submitted data and analysis.

No. of the participant	Thickness of graphene oxide flake		Expanded uncertainty $(k = 2)$
	Mean value/nm	Standard deviation/nm	
1	0.89	0.15	0.31
2	1.05	0.12	0.25
3	0.96	0.22	0.45
4	1.10	0.02	0.13
5	1.25	0.20	0.41
6	1.01	0.02	0.05
7	1.02	0.14	0.28
8	1.04	0.11	0.24
9	1.11	0.30	0.61
10	1.50	0.27	0.56
11	0.52	0.02	0.11
12	1.15	0.36	0.72

If $|E_n| \leq 1$, the difference between the participating laboratory measurement result and the reference value is within reasonable expectation; the comparison results are acceptable.

If $|E_n| > 1$, the difference between the participating laboratory measurement result and the reference value did not reach reasonable expectations, and the reasons should be analyzed.

It is clear from Fig. 5.11 and Table 5.4 that among the measurement results of 12 participating laboratories, E_n absolute values of 11 participating laboratories' results are in the range of less than or equal to 1, so the comparison results of these 11 laboratories are qualified and the discreteness is very small. In addition, the discreteness of one participating lab is large, the main reason is the operator statistical results' error.

5.5.3 International Comparison

Advanced material metrology laboratory of NIM (China) led the international comparison of "Thickness measurement of GO flakes using AFM" under VAMAS technical working group (VAMAS/ TWA41), in order to convene more national metrology institutes with equal qualifications to participate in the comparison. A total of 12 laboratories from metrology institutes/universities participated in this comparison, and the information of the participating laboratories is shown in Table 5.5. The comparison sample and comparison measurement instruction are the same as the domestic comparison; the aim of the comparison is to verify the international equivalence and consistency. The comparison results of GO flake thickness are shown in Fig. 5.12.

The results from Fig. 5.12 show that the thicknesses and associated uncertainty values reported by the participants all overlap

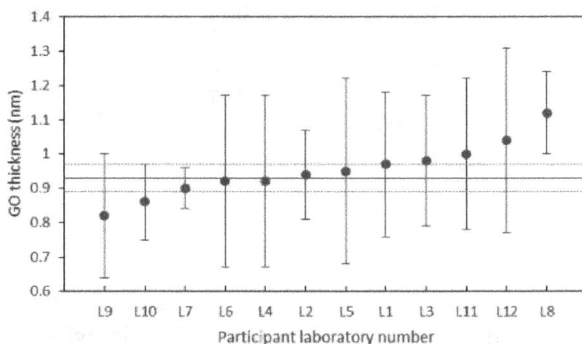

Fig. 5.12.   Average thickness of a GO flake for each participant from the ILC, with their associated uncertainties (u_i). Here, the solid horizontal line represents the CRV (T_{ref}) and the dashed horizontal lines are the standard uncertainty (u_{ref}) of the CRV.

Table 5.5. Information of the participating laboratories.

Country or area	Name of the institution
China	National Institute of Metrology (NIM)
UK	National Physical Laboratory (NPL)
France	National Laboratory of Metrology and Testing (LNE)
Canada	National Research Council (NRC)
Italy	Foundation Bruno Kisler, Italy (CMM MNF)
Australia	The National Measurement Institute (NMIA) of Australia
Brazil	National Agency for Industrial Metrology, standardization and quality of Brazil (INMETRO)
Denmark	National Institute of Metrology of Denmark (DFM)
China Taiwan	*Taiwan* Industrial Technology Research Institute (ITRI)
Australia	Swinburne University of Technology, University of Melbourne
Thailand	National Institute of Metrology of Thailand (NIMT)

with the CRV with its uncertainty, except Participant 8, indicating that the thickness and standard deviation of the samples measured by eleven participating laboratories are within reasonable expectation, the thickness of GO samples provided by laboratories has good equivalence, and further demonstrate the reliability and operability of the method. It is worth mentioning that after discussion and analysis, it may be the inappropriate calibration of AFM is the main reason for Participant 8 out of the CRV interval.[27] Due to the equivalence and consistency of comparison results, the measurement instruction based on the comparison has been approved by the International Organization for Standardization (ISO) and proposes an international standard "ISO/PWI 23879 Nanotechnologies — Structural characterization of GO flakes: thickness and lateral size measurement using AFM and SEM."

In addition, the domestic and international comparison results have also verified that the sample of GO on mica substrate can meet the requirements of homogeneity and stability of CRM. This sample is applied for the national CRM. The development of this CRM can fill the gap of the standard material with the Z-axis displacement value of around 1 nm, boasting scientific implications and practical merits.

5.6   Standardization of Measurement Methods

As is described in Sections 5.2–5.4, by conducting various studies on equipment calibration, sample preparation, sampling and measurement conditions, as well as by verifying through measurement comparison the measurement results to be equivalent and consistent, researchers have confirmed that the reliability, operability and applicability of measurement methods are suitable for scientific research and industrial promotion. Therefore, according to the rules given by GB/T1.1-2009 *Directives for standardization — Part 1: The structure and drafting of standards* and T/CSTM 00002-2019 *Preparation principles of measurement methods standards*, Advanced material metrology laboratory of NIM (China) publishes the group standard T/CSTM 00003-2019 "Thickness measurements of two-dimensional materials Atomic force microscopy (AFM)" through Chinese Society for Testing and Materials (CSTM), Zhongguancun, providing a unified measurement basis for measuring the thickness of graphene-related materials.

This standard has stipulated the measurement principle for two-dimensional material thickness, the requirement for instrument and equipment, sample pretreatment, measurement method, thickness calculation method, uncertainty evaluation of measurement results and measurement reports, etc. It has thus made a unified standardization on the selection of Atomic Force Microscope (AFM) probe, vibration environmental requirements and humidity environmental requirements, instrument calibration, imaging mode, measurement position selection and measurement steps, image smoothing, data selection, data processing and related parameters and requirements, and provided the case of measuring the thickness of GO as a reference. The standard has been used as one of the criteria for the determination of graphene powder materials and cooperates with third-party certification bodies to carry out product testing and verification of graphene powder materials. It has provided a technical basis for reliable and comparable measurement results, and a technical barrier to the elimination of product trade, effectively facilitating the implementation of the whole chain of NQI technology elements in the field of graphene industry and promoting the standardized and healthy development of graphene powder material industry.

5.7 Summary

This chapter has introduced in detail research on measurement technologies of graphene-related material thickness based on AFM technology and its standardization. By elaborating on the establishment, verification and application of calibration and measurement methods within the measurement range (less than 1 nm) of AFM involved in the thickness measurement of graphene-related materials, it has sorted out the trajectory and application of metrological technology of graphene material thickness, hoping to play a reference role for the development of such research work in the future.

References

[1] Ren L, Bu T, Tang Q, *et al.* Investigation of graphene NQI technology. *China Metrology*, 2018, (2): 101–104.

[2] Binnig G, Quate C F, Gerber C, *et al.* Atomic force microscope. *Physical Review Letters*, 1986, 56(9): 930–933.

[3] Cao P G. *Surface Chemistry at the Nanometer Scale.* California: California Institute of Technology, 2011.

[4] Xu K, Cao P G, Heath J R. Graphene visualizes the first water adlayers on mica at ambient conditions. *Science*, 2010, 329(5996): 1188–1191.

[5] Nemes-Incze P, Osvath Z, Kamaras K, *et al.* Anomalies in thickness measurements of graphene and few layer graphite crystals by tapping mode atomic force microscopy. *Carbon*, 2008, 46(11): 1435–1442.

[6] Jung W, Park J, Yoon T, *et al.* Prevention of water permeation by strong adhesion betweengraphene and SiO_2 substrate. *Small*, 2014, 10(9): 1704–1711.

[7] Giusca C E, Panchal V, Munz M, *et al.* Water affinity to epitaxialgraphene: The impact of layer thickness. *Advanced Materials Interfaces*, 2015, 2(16): 1500252.

[8] Eigler S, HofF, Enzelberger-Heim M, *et al.* Statistical Raman microscopy and atomic force microscopy on heterogeneousgraphene obtained after reduction of graphene oxide. *The Journal of Physical Chemistry C*, 2014, 118(14): 7698–7704.

[9] Paredes J I, Villlar-Rodil S, Solís-Fernandez P, *et al.* Atomic force and scanning Tunneling microscopy imaging of graphene nanosheets derived from graphite oxide. *Langmuir*, 2009, 25(10): 5957–5968.

[10] Jalili R, Aboutalebi S H, Esrafilzadeh D, *et al.* Organic solvent-based graphene oxide liquid crystals: A facile route toward the next generation of self-assembled Layer-by-layer multifunctional 3D architectures. *ASC Nano*, 2013, 7(5): 3981–3990.

[11] Solís-Fernandez P, Paredes J I, Villlar-Rodil S, *et al.* Determining the thickness of Chemically modifiedgraphenes by scanning probe microscopy. *Carbon*, 2010, 48(9): 2657–2660.

[12] Schmidt U, Dieing T, Ibach W, *et al.* A confocal Raman-AFM study of graphene. *Microscopy Today*, 2011, 19(6): 30–33.

[13] Nagisa H, Makoto T, Masaru T. Effect of laser irradiation on few-layergraphene in air probed by Ramaptscopy. *Transactions of the Materials Research Society of Japan*, 2013, 38(4): 579–583.

[14] Novoselov K S, Jiang D, Schedin F, *et al.* Two-dimensional atomic crystals. *Proceedings of the National Academy of Sciences of the United States of America*, 2005, 102(30): 10451–10453.

[15] Novoselov K S, Geim A K, Morozov S V, *et al.* Electric field effect in atomically thin carbon films. *Science*, 2004, 306(5696): 66–69.

[16] Obraztsova E A, Osadchu A V, Obraztsova E D, *et al.* Statistical analysis of atomic force microscopy and Raman spectroscopy data for estimation of graphene Layer numbers. *Physica Status Solidi (b)*, 2008, 245(10): 2055–2059.

[17] Ochedowski O, Begall G, Scheuschner N, *et al.* Graphene on Si (111) 7 × 7. *Nanothlogy*, 2012, 23(40): 405708.

[18] Ishigami M, Chen J H, Culllen W G, *et al.* Atomic structure of graphene on SiO_2. *Nano Letters*, 2007, 7(6): 1643–1648.

[19] Bienias M, Hasche K, Seeemann R, *et al.* Metrological atomic force microscope. *Journal of Metrology*, 1998, 19(1): 1–8.

[20] Gao S. *A Study of the Metrological Photomicroscope.* Tianjin: Tianjin University, 2007.

[21] JJF 1351–2012 Calibration Specification for Scanning Probe Microscope [S]. (Technical Specification) Released by National Metrology Technical Specifications of the People's Republic of China

[22] Mechler Kokavecz J, Heszler P, *et al.* Surface energy maps of nanostructures: Atomic force microscopy and numerical simulation study. *Applied Physics Letters*, 2003, 82(21): 3740–3742.

[23] Shear C J, Slatttery A D, Stapleton A J, *et al.* Accurate thickness measurement of graphene. *Nanotechnology*, 2016, 27(12): 150704.

[24] Blake C, *et al.* Makinggraphene visible. *Applied Physics Letters*, 2007, 91(6): 063124.

[25] Lui C H, Liu L, Mak K F, *et al.* Ultra flatgraphene. *Nature*, 2009, 462(7271): 339–341.

[26] Yao Y X, Ren L L, Gao S T, *et al.* Histogram method for reliable thickness measurements of graphene films using atomic force microscopy (AFM). *Journal of Materials Science & Technology*, 2017, 33(8): 815–820.

[27] Bu T J, Gao H F, Yao Y X, *et al.* Thickness measurements of graphene oxide flakes using atomic force microscopy: Results of an international interlaboratory comparison. *Nanotechnology*, 2023, 34(22): 225702.

Chapter 6

Metrological Technology of Electron Microscope for Graphene-Related Materials

Xu Li, Guocai Dong, Xiaomin Zhang, and Jinlong Li

With a very complete crystal structure and superior stability, graphene boasts exceptional electrical, mechanical, optical and thermal properties and ultra-high specific surface area, showing great application potential in electronic devices, sensors, energy storage materials and other fields. With the development of research, studies on the micro-morphology, number of layers, lattice spacing, atomic structure and defects of graphene have become more and more important. Electron microscopy technologies include scanning electron microscope (SEM) and transmission electron microscope (TEM), one of the important means to characterize the morphology, structure and composition of materials. The electron microscope can be used not only to obtain the morphology, diffraction spectrum, high-resolution lattice images and atomic images of graphene but also to accurately measure the number of graphene layers and the lattice spacing, hence it is an important tool for scientific research and qualitative judgment of graphene-related materials. But in practical applications, for both the SEM and the TEM, there is a measurement deviation in the range of values to which they apply. For example, when SEM is subject to the influence of such factors as the filament lifetime, voltage stability and current stability, the magnification of electron microscope will deviate with the extension of using time.[1] The quality of

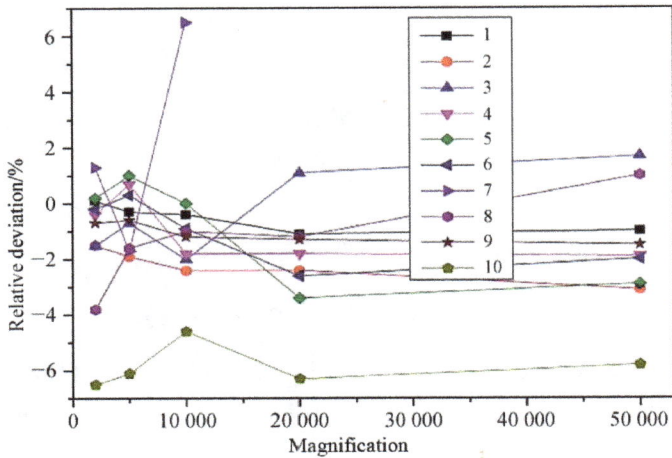

Fig. 6.1.   Results of 10 sets of SEM measurements of one-dimensional grid reference.

the SEM image is influenced by the path of the scanning electron, which is likely to cause image magnification deviation, horizontal and vertical image distortion, inaccurate size measurement and other problems.[2] Figure 6.1 shows the results of 10 sets of SEM measurements of one-dimensional grid reference. It is clear from the figure that there are great differences in the measurement results from different laboratories, with the maximum deviation reaching 11.1%.[3] Likewise, the measurement results of TEM are also subject to the influence of factors such as the filament life, acceleration voltage, current stability, magnetic field stability and sample quality during its use; the magnification of electron microscope will deviate with the extension of using time. Figure 6.2 shows the results of 9 sets of TEM measurements of {220} crystal plane distance at high magnification. It is clear from the figure that the measurement results of the same sample by different TEMs have great difference; the maximum deviation is 8.81%. Thus, the results measured by different SEMs and TEMs are quite different, so the measurement results of different laboratories are not consistent. Metrological technology of sheet size, number of layers and crystal plane distance of graphene materials are the very research to make the measurement results consistent and comparable.

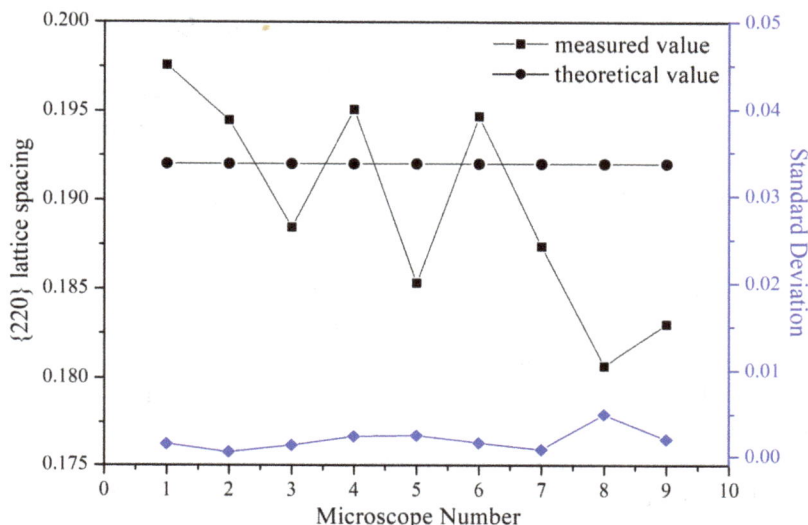

Fig. 6.2.   Results of nine sets of TEM measurements of {220} crystal plane distance at high magnification.

According to the connotation and category of materials metrology, metrological technologies for graphene-related materials under the electron microscope include electron microscopy calibration traceability and measurement method applications. This chapter introduces these two aspects in detail.

6.1   Measurement of Sheet Size and Coverage of Graphene-Related Materials by SEM

SEM is a multi-functional device for characterizing the morphology, structure and composition of materials. SEM has the advantages of high resolution, wide amplification range, simple sample preparation process and depth of field. It is widely used in the fields of new materials, semiconductors, microelectronics and life science, playing a very important role in morphology characterization, structure measurement and composition analysis. SEM can directly measure the information on grain morphology, grain size, coverage range, nucleation density and growth rate of graphene. SEM is usually equipped with several detectors, like electron energy loss analysis probe, which can analyze the chemical composition of graphene. Some of them are equipped with an electron backscatter diffraction detector, which can be used to determine the crystal orientation of the substrate, and then study the relationship between the shape of the grains and the orientation of the substrate. SEM works by scanning the surface of a sample using a convergent electron beam. The electron beam interacts with the sample to produce secondary electrons, backscattered electrons, Auger electrons, characteristic X-ray, etc. By collecting these electrons or rays, the surface morphology, structure and composition of the sample can be obtained. Spatial resolution of SEM depends on the beam diameter and the volume of the beam interacting with the sample; at present, the spatial resolution of many field emission SEMs can achieve 1 nm or so. As a very important means of micro- and nanosize measurement, SEM is widely used in scientific research and practical production. Therefore, in order to ensure the accuracy and reliability of SEM measurement results, it is necessary first to periodically conduct tracing calibration of SEM, so as to ensure that the dimensional measurements are traceable to the definition of meter in the International System (SI).

6.1.1   Traceability and Calibration of SEM

Using SEM to measure the sheet dimensions of graphene-related materials, the first step is to make sure that the SEM image and scale are accurate. Therefore, it is necessary to calibrate the SEM. The premise of SEM calibration is to study the tracing technique of SEM, establish the tracing path of the measurand value of SEM and formulate the calibration criterion of SEM, so as to provide calibration or calibration service for the end user. Figure 6.3 is the tracing path of SEM. It is clear from the figure that assigning values to certified reference materials (CRMs) by using traceability to meter-defined standard apparatus for nanoscale geometric structures in SI ([2007] No. 40 of National Social Science Standard Metrology Certificate), CRMs are carriers of accurate value transmission. Through CRMs, characteristic parameters of SEM are calibrated according to the calibration specifications, the uncertainty of calibration result is evaluated, and the accurate value of traceability to SI basic unit is transmitted to the end user. Therefore, let's introduce the CRMs related to SEM first.

SI length standard

↑

Laser wavelength

↑

Standard device for nanoscale geometric structures

↑

Grid reference material/ particle size reference material

↑

End-user's SEM

↑

Size value of the measured sample

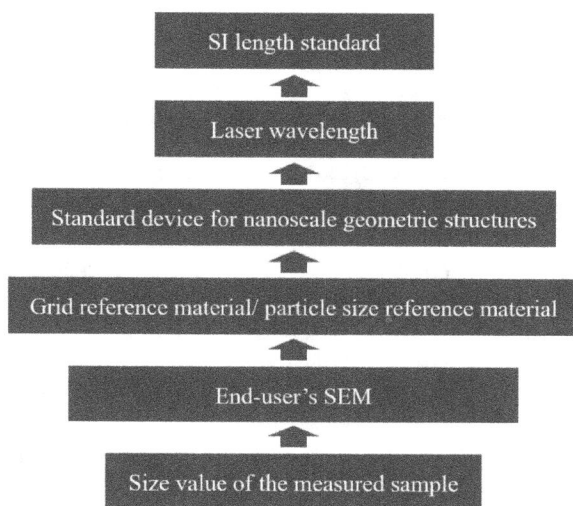

Fig. 6.3. Traceability path of SEM.

6.1.1.1 *CRMs*

The calibration of SEM requires certified CRMs. China has developed a series of materials for SEM magnification, rate calibration, contamination rate or drift rate calibration. National Sharing Platform for Reference Materials (http:www.ncrm.org.cn) gives information on existing CRMs in China. Take CRMs for magnification calibration of SEM as an example, there are mainly one-dimensional and two-dimensional grids, line widths and line distance template polystyrene balls, gold particles, etc. Table 6.1 lists some CRMs for magnification calibration of SEM developed in China, and Fig. 6.4 is the images of CRMs for SEM calibration developed at home and abroad. When the magnification of SEM is calibrated, the CRMs with equivalent nominal values can be selected to carry out calibration according to the range of values to be measured.

Many CRMs for SEM calibration have been developed abroad as well. US National Institute of Standards and Technology (NIST) developed polystyrene spheres [SRM1964, Fig. 6.4(e)], two-dimensional lattice of monocrystalline silicon [RM 8820, Fig. 6.4(d)], gold particles (RM8011), etc. Physikalisch-Technische Bundesanstalt (PTB) developed silicon dioxide, a thin film thickness reference material (IMS-HR 94 175-04). The National Physical Laboratory (NPL)

Table 6.1. Some CRMs for magnification calibration of SEM developed by China.

Name	No.	Reference value/nm	Development unit
Gold particle size reference material of nanometer grade	GBW(E)120126 GBW(E)120127 GBW(E)120150	22.6 43.7 11.8	National Center for Nanoscience
Particle size reference material of polystyrene microspheres	GBW12009 GBW12010 GBW(E)120062 GBW(E)120063	855 333 582 299	China University of Petroleum
One-dimensional nanogrid CRMs used for scanning probe microscopy and SEM	GBW13956	400.5	Institute of Physics, Chinese Academy of Sciences

of UK developed the reference material for metal grating. Russian Certification Center (GOST-R) developed the reference material of metal periodic stripe structure (CRM 6261-91). In addition, some commercial companies also developed and sold calibration samples for SEM. British science company Agar Science sold reference grid samples (2,160 line/mm). US Ted Pella Company sold carbon replica parallel, mesh grid and single crystal silicon mesh grid (10 μm), US Advanced Surface Microscopy, Inc. sold one-dimensional nanowire spacer templates. US Geller company sold MRS-4 series of multi-functional standard samples, etc.

6.1.1.2 *Calibration of SEM*

In 1997, State Education Commission of the PRC published JJG 010-1996 *Verification Regulation of Analytical Scanning and SEM*. This protocol provides a clear classification of analytical scanning electron microscopy, specifies verification requirements for the indication error of electron microscope magnification, repeatability of magnification indication and resolution of secondary electronic images and adds the important accessories of SEM — technical index and verification method of X-ray energy spectrometer. Although

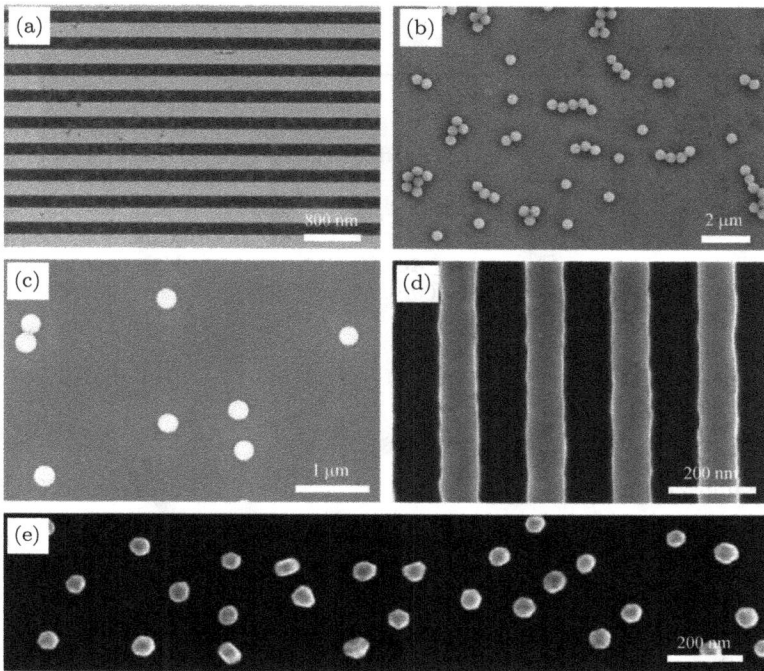

Fig. 6.4. Images of some CRMs for SEM calibration developed at home and abroad: (a) one-dimensional nanogrid reference material made in China (GBW13956, reference value is 400.5 nm); (b,c) CRMs for particle size polystyrene microspheres made in China [GBW(E) 120062 and GBW(E) 120063, reference values are 582 nm and 299 nm, respectively]; (d) two-dimensional grid line distance reference material for monocrystalline silicon made by NIST (RM8820, reference value is 200 nm); (e) CRM for particle size polystyrene microspheres made by NIST (SRM 1964, reference value is 60 nm).

this protocol presents a fairly scientific and effective evaluation of important indicators of SEM metrological performance, it was published and implemented more than two decades ago when China had not developed relevant CRMs; the verification regulation of SEM could not thus be effectively carried out. With the steady progress of science and technology and the rapid development of electron microscopy, in 2011, the General Administration of Quality Supervision, Inspection and Quarantine published GB/T27788-2011 *Microbeam Analysis-Scanning Electron Microscopy-Guideline for Image Magnification Calibration*. This guideline only specifies the

Table 6.2. Main metrological characteristics of calibration and requirements for CRMs.

Measurement characteristics	Magnification (M)	Expanded uncertainty of CRMs ($k = 2$)
Magnification error	$70{,}000 < \text{M} \leq 200{,}000$	Less than 5%
	$10{,}000 < \text{M} \leq 70{,}000$	Less than 6%
	$\text{M} \leq 10{,}000$	Less than 8%
Repeatability of magnification indication	$10{,}000 \leq \text{M} \leq 70{,}000$	Less than 6%
Linear distortion of image	$500 \leq \text{M} \leq 2{,}000$	Less than 6%
Sample contamination rate	$10{,}000 \leq \text{M} \leq 70{,}000$	Less than 6%
Sample drift rate	$10{,}000 \leq \text{M} \leq 70{,}000$	Less than 6%

most important measurement parameter for SEM calibration — the calibration of magnification, without mentioning the calibration of linear distortion, sample contamination rate and sample drift rate. China National Institute of Metrology has developed metrological SEM, and on this basis, it has issued *Specification* for *SEM Calibration* (submitted for approval). The following is an introduction to the calibration characteristics of SEM and calibration methods. The main metrological characteristics of calibration and requirements for CRMs are shown in Table 6.2.

6.1.1.3 *Magnification calibration*

It is necessary to calibrate magnification since magnification is the representative parameter of the accuracy of SEM. The magnification of an SEM is the ratio of the image size to the actual size of the object. The magnification M is given by Equation (6.1):

$$M = \frac{L_{\text{image}}}{L_{\text{ture}}} \qquad (6.1)$$

where L_{image} is the measurement of the grid distance on a film, photograph or charge-coupled device (CCD) camera after the CRM is enlarged and imaged by SEM and L_{ture} is the reference value of CRMs.

Today's SEM is equipped with large-scale, high-resolution CCD cameras; the measurement results are all saved and output in electronic images. The scale in the image is the correlation between the reference material object and the reference material image and calculates the unique display value for the magnification of the image. Therefore, the magnification of the calibrated SEM is the calibration of the scale to which the electronic image corresponds at this magnification.

Let us take the one-dimensional nanogrid CRM (GBW13956) made in China, for example, an SEM image is taken of the CRM with a scale (Fig. 6.5). In this electronic image, the 400 nm scale in the lower right corner is the characteristic length consisting of a number of pixels, while the grid distance D_m is also the characteristic length of a number of pixels, and the unit pixel size of the two is equal. Therefore, the calibration of electronic image scale is obtained by dividing the reference value of the unit grid distance by the actual measured value of the unit grid distance. The measurand value of the unit grid distance in the reference material can be measured by the built-in software for EM or ImageJ software. Calculate the indication error and calibration coefficient of the scale according

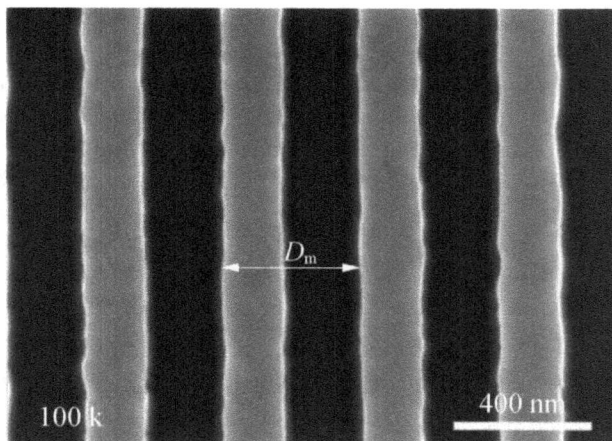

Fig. 6.5. SEM image of one-dimensional nanogrid reference material (GBW13956) made in China.

to Equations (6.2) and (6.3), i.e. the magnification of the indicator error and calibration coefficient:

$$\Delta_s = \frac{D_m - D_c}{D_c} \tag{6.2}$$

$$\eta = \frac{D_c}{D_m} \tag{6.3}$$

where Δ_s is the indication error of the scale, η is the calibration factor of the scale, D_m is the measurement of the unit grid distance in the reference material, and D_c is the reference value of the unit grid distance in the reference material.

6.1.1.4 *Repeatability of magnification indication*

In order to test the stability of the instrument, the repeatability of indication shall be calibrated. Select the appropriate reference material and the appropriate magnification according to the actual use of SEM. Take the first image of the reference material as per normal operating procedure, change the acceleration voltage of the electron beam, revert back to the setting of the first photo five minutes later and then take a second image of the reference material. Repeat the procedure until 10 images are taken in 45 minutes. Analyze the electronic image of reference material obtained by ImageJ software, calculate the number of pixels and the average number of pixels contained in the reference material feature size on the 10 images, and use pixel number instead of the actual measurand value of characteristic size to calculate magnification rate indication repeatability. The equation for calculating the repeatability of magnification indication g is

$$g = \frac{3\sigma}{\overline{P}} = \frac{3 \times \sqrt{\frac{\sum_{i=1}^{n}(P_i-\overline{P})^2}{n}}}{\overline{P}} \tag{6.4}$$

where P_i is the number of pixels contained in the characteristic size of the reference material in the ith image ($i = 1, 2, 3, \ldots, 10$), P is the average number of pixels contained in the characteristic size of the reference material, n is the number of electronic images ($n = 10$) and σ is the total standard deviation of pixels. As is shown in Fig. 6.6, select the unit grid distance of the reference material as the object

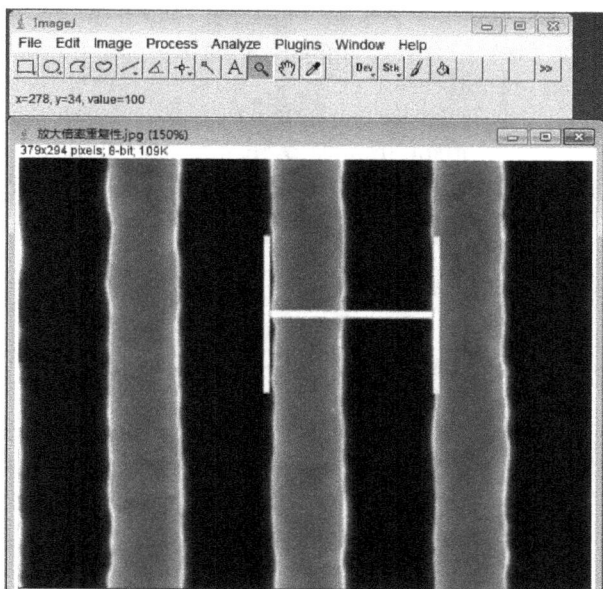

Fig. 6.6.   Magnification indication reproducibility of SEM using ImageJ software and reference material pixel number.

Table 6.3.   Calculation of the repeatability of SEM magnification by using pixel number.

Number of pixels contained in the grid distance	Serial number of electronic images										$\bar{p}$	3σ	g
	1	2	3	4	5	6	7	8	9	10			
	108	107	106	108	109	107	107	106	108	106	107.2	3.1	2.89%

of investigation, take 10 electronic photos at 20,000 magnifications and measure the number of pixels contained in the grid distance by ImageJ software; the result is shown in Table 6.3. According to Equation (6.4), when the magnification of 20,000× is calculated, the repeatability of magnification indicated by SEM is 2.89%.

6.1.1.5   *Linear distortion of images*

SEM image distortion is very serious at low magnification; therefore, it is necessary to calibrate the image linear distortion in the range of low magnification according to actual use. Place polystyrene ball reference materials in the center and corners of the display screen, take a total of five electronic images, measure the number of pixels contained in the diameter of polystyrene sphere in directions x and y in five images respectively by ImageJ software and then calculate the linear distortion of the images by Equations (6.5) and (6.6). The linear distortion of x-direction image is Δ_x and the linear distortion of y-direction image is Δ_y; there is

$$\Delta_x = \frac{\Delta P_{x,\max}}{P_{x0}} \times 100\% \tag{6.5}$$

$$\Delta_y = \frac{\Delta P_{y,\max}}{P_{y0}} \times 100\% \tag{6.6}$$

where P_{x0} is the number of pixels contained in the diameter of the polystyrene sphere in direction x at the central position of the image, P_{xi} is the number of pixels contained in the diameter of the polystyrene sphere in direction x at the i position of the image $(i = 1, 2, 3, 4)$, $\Delta P_{x,\max}$ is the maximum of $|\Delta P_{xi}|$, $\Delta P_{xi} = P_x - P_{x0}$, P_{y0} is the number of pixels contained in the diameter of the polystyrene sphere in the direction y at the central position, P_{yi} is the number of pixels contained in the diameter of the polystyrene sphere in direction y at the i position of the image $(i = 1, 2, 3, 4)$ and $\Delta P_{y,\max}$ is the maximum of $|\Delta P_{yi}|$, $\Delta P_{yi} = P_{yi} - P_{y0}$.

As is shown in Fig. 6.7, calculate the linear distortion of the image with the same polystyrene ball diameter in directions x and y at 500× magnification and measure the number of pixels of the same polystyrene sphere diameter in directions x and y of the display screen using ImageJ software, and the results show (Table 6.4) that the linear distortion of SEM electron image in direction x under 500× magnification is 1.9%, and the linear distortion of SEM electron image in direction y is 2.0%.

6.1.1.6   *Sample contamination rate and sample drift rate*

In the use of SEM, the samples containing C, H and O elements are subjected to long-term electron beam irradiation in the microscope, producing volatile pollutants which polymerize to form carbon deposits, resulting in a black square or black blob in the sample imaging area. This will either block the aperture or contaminate the electron microscope probe and other components, thus leading

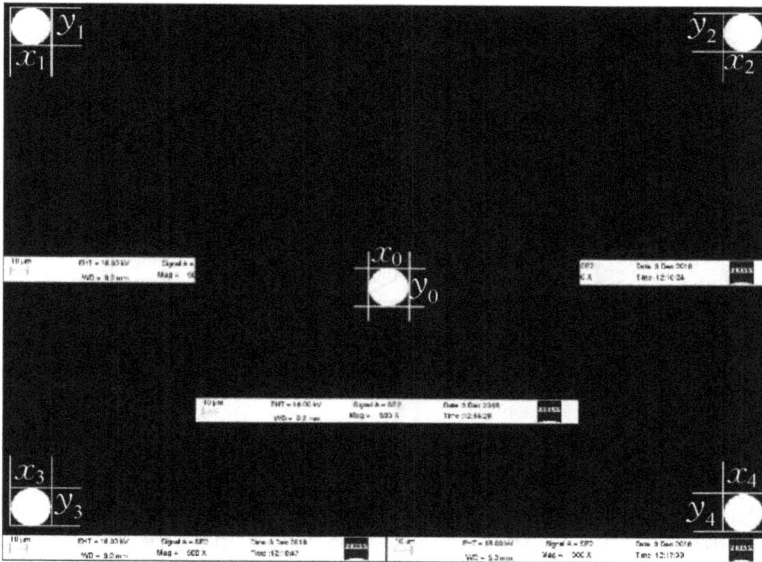

Fig. 6.7.   Electronic image of the same polystyrene ball in different positions on the display screen.

Table 6.4.   Calculation of the linear distortion of SEM image in directions x and y by using pixel number.

	Number of pixels in the diameter of the polystyrene sphere P_i						
Direction	Position 0	Position 1	Position 2	Position 3	Position 4	ΔP_{max}	$\frac{\Delta P_{max}}{P_0}/\%$
x	105	106	106	107	105	2	1.9
y	101	100	102	100	102	2	2.0

to problems with the reliability, accuracy and safety of SEM measurements. SEM requires high stability at high magnification and extremely low drift allowance. Factors causing SEM drift include deformation and movement of samples due to ambient vibration, magnetic field, noise and electron beam bombardment. If the SEM were to drift, it would cause the image to blur and the image movement and the measurement data accuracy to deteriorate. At present, the number of SEMs in China is over 3,000 and is increasing at the rate of over 200 per year; the needs for SEM verification, calibration and maintenance are huge. The contamination rate and drift rate of

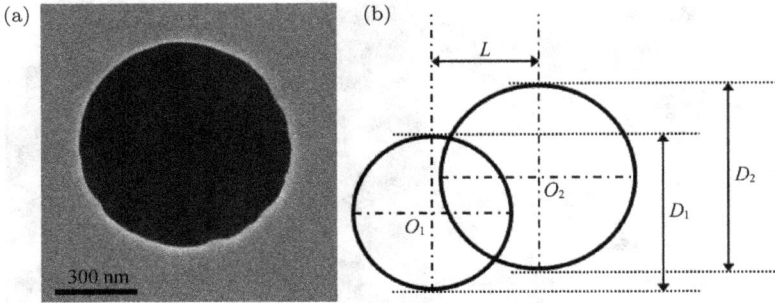

Fig. 6.8.   Schematic figure of SEM image of microporous carbon film reference material as well as sample contamination rate and sample drift rate measurement.

SEM require periodic testing and calibration using CRMs to support the periodic maintenance of SEM.

Adjust SEM to the best focusing state and calibrate with microporous carbon film reference material. Select a micropore and move it to the center of the display screen, enlarge the image till the area of the round hole is about 80% the area of the display screen, take the first image at this magnification with all parameters unchanged and then take the second image after 10 minutes, as is shown in Fig. 6.8. Calculate the sample contamination rate and the sample drift and shift rate according to Equations (6.7) and (6.8):

$$Q_C = \frac{D_2 - D_1}{2t} \tag{6.7}$$

$$Q_D = \frac{L}{t} \tag{6.8}$$

where Q_C and Q_D are the rate of sample contamination and the rate of sample drift, respectively, t is the time interval between the two takes, which is normally set at 10 minutes, D_1 and D_2 are the micropore diameters measured on the images taken before and after the interval t and L is the distance from the center to the micropores D_1 and D_2.

6.1.1.7   Uncertainty evaluation of calibration results

6.1.1.7.1   Mathematical models

According to the standard of measurement methods, the characteristic size in the sample is measured using a calibrated SEM.

The measurand value of the characteristic size may be subject to the influence of noise, magnetic field, vibration, voltage, magnetic lens astigmatism, sample drift and other factors, as well as CRMs and measurement methods. The calibration value and calibration coefficient of the measurand value of the characteristic size are calculated according to Equations (6.9) and (6.10):

$$S_{mc} = S_m \times \eta \qquad (6.9)$$

$$\eta = S_0/S \qquad (6.10)$$

where S_{mc} is the calibration value of the measurand value of the characteristic size measured by SEM, S_m is the measurand value of the characteristic size measured by SEM, η is the calibration factor for the magnification (or pixel size) used and S_0 is the reference value of CRMs for SEM calibration.

Calculate the calibration value S_{mc} for measurand value of the characteristic size according to Equations (6.9) and (6.10), and further calculate the uncertainty of calibration value $u(S_{mc})$ according to Equation (6.11):

$$u(S_{mc}) = \sqrt{c_1^2 u(S_m)^2 + c_2^2 u(S_0)^2 + c_3^2 u(S)^2} \qquad (6.11)$$

where $c_1 = \frac{\partial S_{mc}}{\partial S_m} = \frac{S_0}{S}$, $c_2 = \frac{\partial S_{mc}}{\partial S_0} = \frac{S_m}{S}$, $c_3 = \frac{\partial S_{mc}}{\partial S} = -\frac{S_m S_0}{S^2}$.

6.1.1.7.2 Evaluation of A-type uncertainty

When using the reference material to calibrate SEM, select a suitable location in the reference material, repeat the measurements of the characteristic value of reference materials at least six times and then calculate according to Equation (6.12) the uncertainty introduced by the repeatability of the reference material measurement $u(S)$:

$$u(S) = \sqrt{\frac{1}{n-1} \sum_{j=1}^{n} (x_j - \bar{x})^2} \qquad (6.12)$$

where n is the total number of measurements at the selected location, j is the serial number for the jth measurement, $j = 1, 2, \ldots, n$, x_j is the measurand value at the jth measurement and $\bar{x}$ is the average of all the measurements.

6.1.1.7.3   Evaluation of B-type uncertainty

The uncertainty of the reference material used for the magnification calibration of SEM is U_{CRM} ($k = 2$), therefore, the uncertainty introduced by the reference material for the magnification calibration of SEM is $u(S_0) = U_{\mathrm{CRM}}/2$.

6.1.1.7.4   Combined uncertainty

Calculate the combined uncertainty $u(S_{\mathrm{mc}})$ according to Equation (6.11).

6.1.1.7.5   Expanded uncertainty

The calibration value of the characteristic size of the sample to be measured obeys normal distribution, the confidence level is 95%, and expansion factor $k = 2$, therefore, the Expanded uncertainty U for the calibration value of the characteristic size measurand value of the sample to be measured is

$$U = 2 \times u(S_{\mathrm{mc}}) \tag{6.13}$$

6.1.2   Measurement Methods of Graphene Sheet Size

Dispersion, shape, size and distribution of graphene powders in solvents are vital to improving the performance of graphene-related materials, such as electrical conductivity, thermal conductivity, intensity, hardness and so on. Compared with the traditional sample which becomes spherical or linear-like when dispersed in the dispersion solution, graphene powder materials are irregular sheets in solution, and the size in x- and y-axis directions is much larger than that in the z-axis direction. This makes dynamic light scattering, laser particle size and other instruments unsuitable for measuring the size and shape of such materials. SEM can directly observe the shape and size of graphene materials, but the parameters by which their shape, size and distribution can be described have to be settled immediately. The author has carried out preliminary research on these issues. The following is an introduction to the research proposal and some findings, which can serve as a reference for other researchers.

6.1.2.1   *Sample preparation*

Weigh 0.002 g graphene oxide powder with an analytical balance and feed it into a centrifuge tube of 15 ml. Fill the centrifuge tube with anhydrous ethanol and cover it, oscillate the tube with a vortex oscillator for 10 minutes to obtain the dispersion liquid. Then put the dispersion liquid into an ultrasonic cleaning instrument for ultrasonic dispersion of 3 minutes. Grow a layer of gold thin film of about 5 nm thick on a single crystal and silicon wafer by sputtering deposition method and then drop the graphene oxide dispersion onto a single crystal silicon wafer with a thin film of gold. Obtain the dispersed graphene oxide sample when it dries naturally.

6.1.2.2   *Measurement process*

Calibrate SEM according to Section 6.1.1. Place the prepared graphene oxide sample into a calibrated SEM, select an appropriate acceleration voltage and working distance and take SEM images of the graphene oxide sample; the number of pixels of the electronic image at different magnifications is different. For example, for an image of 2,048 pixels × 1,536 pixels, the scale 2 μm contains 105 pixels and the unit pixel size is (2/105) micron/pixel; for an image of 3,072 pixels × 2,304 pixels, the scale 2 μm contains 159 pixels and the unit pixel size is (2/159) micron/pixel. On this basis, the size of each graphene sheet can be calculated. As can be seen from Fig. 6.9(a), the oxide graphene sheet has good dispersion and a clear shape.

6.1.2.3   *Data analysis*

It is clear from Fig. 6.9(a) that graphene sheets are in the shape of bars, squares, sub-circles and other irregular shapes. Although the diameter and equivalent diameter parameters can be used to describe circular and quasi-circular shapes and sizes, more suitable parameters are needed to describe and define the morphology, size and distribution of graphene sheets and provide data analysis methods since the graphene lamellae have a two-dimensional structure and a complex and variable shape and size.

At present, the research is in progress and has achieved some results. Suitable parameters for describing the shape and size of

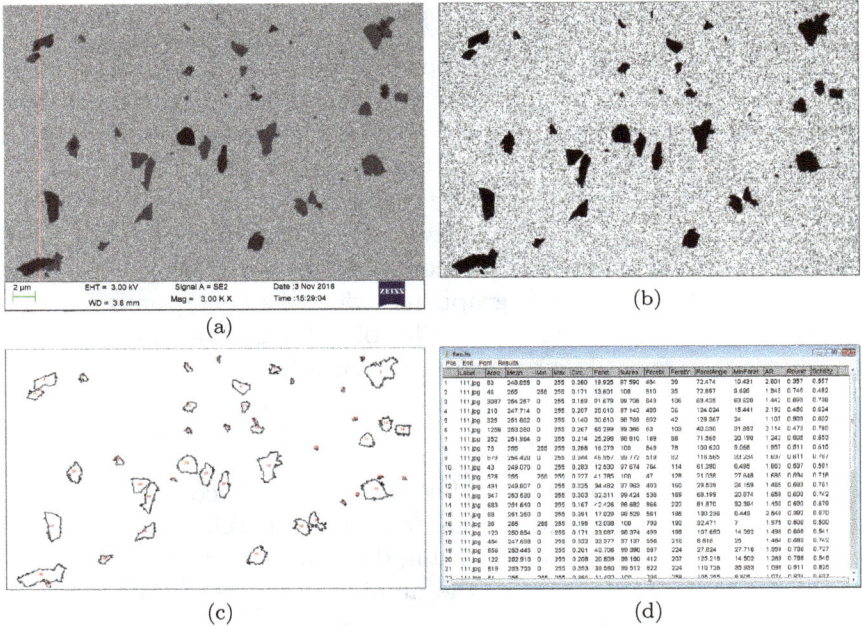

Fig. 6.9.   Measurement process and results of SEM image and size calculation of graphene oxide lamellae: (a) SEM image of graphene oxide lamellae; (b) image contrast processing; (c) identification of graphene oxide lamellae images by ImageJ software; (d) calculation results of the size of graphene oxide lamellae.

two-dimensional lamellae are screened based on the parameter description of particle shape and size.[4] The requirements for the description parameters are as follows: One is to have as few parameters as possible so that the user can fully understand the shape, size and distribution of the description, and the other is to have parameters convenient for analysis and calculation. Table 6.5 introduces the parameters of a lamellar graph. Analysis of 12 graphic parameters shows that the area can be used to express the size of two-dimensional lamellar size; the degree of squareness and roundness can be used to express the degree to which the lamellar is close to a circular or square; it is a description of the shape. These three parameters are the most basic parameters; the others will be collected through the continued study of user comments, to fully describe the size, shape and distribution of the two-dimensional material.

Table 6.5.   Parameters that describe a lamellar graph.

Serial number	Name	Symbol	Description	Graphic expression	Equation
1	Area	A	Actual number of pixels in a graph		
2	Perimeter	p	Total length of the image edge		
3	Centroid X	CX	Horizontal coordinate of the centroid		$\Sigma X/A$ of which X is the horizontal coordinate of one point
4	Centroid Y	CY	Vertical coordinate of the centroid		$\Sigma Y/A$ of which Y is the vertical coordinate of one point
5	Length/pixel of the bounding box	BL	Length of the smallest rectangle parallel to the axis of coordinates		$X_{max} - X_{min}$ (X_{max}/X_{min} is the largest or smallest of all X points)
6	Width/pixel of the bounding box	BW	Width of the smallest rectangle parallel to the axis of coordinates		$Y_{max} - Y_{min}$ (Y_{max}/Y_{min} is the largest or smallest of all Y points)

(*Continued*)

Table 6.5.   (*Continued*)

Serial number	Name	Symbol	Description	Graphic expression	Equation
7	Degree of squareness	Extent	Number of pixels in a graph in proportion to the total number of pixels in its corresponding bounding box. For a complete square, its range value is 1		$A/(\mathrm{BL} \times \mathrm{BW})$
8	Roundness	C	Degree of roundness of a graph. For a complete circle, its roundness value is 1	/	$\dfrac{4\pi A}{P^2}$
9	Solidity	Solidity	Ratio of the number of pixels in a graph to the total number of pixels in its corresponding convex hull		$\dfrac{A}{A_{\text{convex hull}}}$
10	Maximum Ferrette diameter/pixel	XF_{max}	Maximum distance between two parallel tangents of a graph		/
11	Minimum Ferrette diameter/pixel	XF_{min}	Minimum distance between two parallel tangents of a graph		/
12	Equivalent diameter	X_p	Diameter of a circle of the same area as an irregular graph		$\sqrt{\dfrac{4A}{\pi}}$

6.1.2.4 *Image processing*

The first step is to crop the image, cutting out the scale and other marks in the electronic image, because these scales and marks will affect the analysis of image contrast and graphene sheet size. About 10% of the scale image in the original image in Fig. 6.10(a) needs to be cropped out; the remaining 90% or so image is shown in Fig. 6.10(b). Then, in order to remove the noise in the image, the image needs to be processed by Gauss filtering. Gauss filter is a linear smooth filter which processes the whole image by weighted average method to eliminate Gauss noise. The Gauss filter value for each pixel is obtained by a weighted average of the initial pixel number of the point and the number of surrounding pixels.

After the image is filtered by Gauss, the binarization process needs to be continued, the purpose is to process the image into contrast,

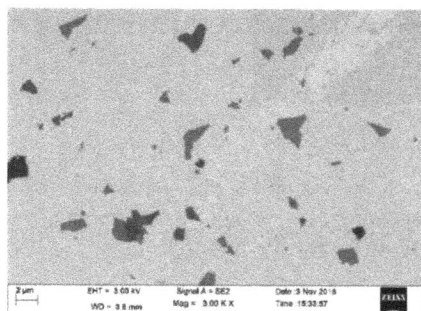

(a)

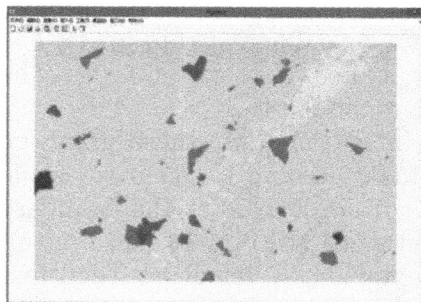

(b)

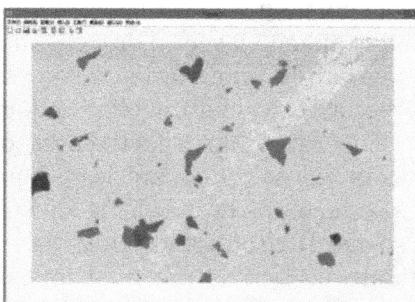

(c)

Fig. 6.10. Graphene sheet images at different stages of processing: (a) original image; (b) cropped image; (c) Gauss-filtered image.

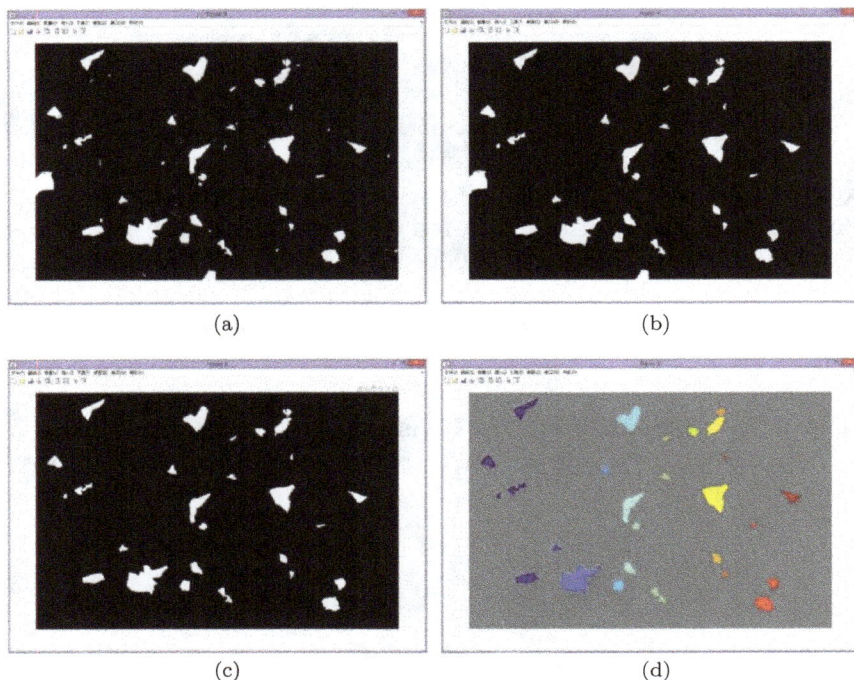

Fig. 6.11. Graphene sheet images at different stages of processing: (a) image after binarization processing; (b) image after removing small slices; (c) image after removing incomplete slices; (d) final image.

clear black and white. Binary processing is based on the maximum intra-group variance of pixels to distinguish the background and graphene layers. Different images have different thresholds. The image after binarization processing is shown in Fig. 6.11(a). The foreground is white, while the background is black.

To ensure the effectiveness of image analysis, those graphene sheets whose pixels are less than 1,000 need to be removed, while those incomplete graphene sheets that are located at the boundary also need to be removed, as is shown in Figs. 6.11(b) and 6.11(c). Figure 6.11(d) is the final image obtained by removing small and incomplete graphene sheets.

After analyzing 576 graphene lamellae in 22 SEM images, the obtained data analysis results are shown in Table 6.6. The average

Table 6.6. Results of data analysis of graphene lamellae.

Parameter	Area/μm^2	Permeter/μm^2	Length/Pixel of bounding box	Width/Pixel of bounding box
Mean value	1.963359	6.865847	1.724324	1.718781
Variance	5.796775	27.67697	1.360846	1.206154
Standard deviation	2.407649	5.26089	1.166553	1.09825

Parameter	Range/degree of squareness	Roundenss	Solidity	Equivalent diameter/μm
Mean value	0.528885	0.474626	0.812695	1.354784
Variance	0.012097	0.027558	0.011556	0.646476
Standard deviation	0.109987	0.166005	0.1075	0.804037

area of these graphene sheets is about 1.963 4 μm^2, and the mean degree of squareness and roundness are both about 0.5, which indicates that the graphene sheets are far from circular or rectangular in shape. Thus, it can be seen that the three parameters of area, squareness, and roundness are sufficient to describe the size and shape of a two-dimensional slice.

6.1.3 Measurement Methods for Coverage of Graphene Films on a Wide Range of Metal Substrates

Due to its exceptional physical properties, graphene has broad application prospects in many fields, such as electronics, photoelectronics and biosensing.[5,6] Compared with graphene obtained by mechanical stripping,[7,8] oxidation–reduction[9] or SiC epitaxial growth,[10,11] graphene grown on a metal substrate has significant advantages.[12] By providing a large-sized metal substrate, a corresponding sized graphene film can be obtained, and there are methods to easily transfer graphene films onto other substrates. For graphene films grown on copper or nickel using chemical vapor deposition,[13, 14] their advantage of convenient transfer is more apparent. The transferred graphene films have high light transmittance and high conductivity, so they can be made into transparent electrodes[15, 16] widely used in all kinds of flexible optoelectronic devices, including touch screen sensors, organic light-emitting diode and organic photovoltaic devices. The quality of graphene films themselves is one of the performance indicators of devices.

At present, transmittance and square resistance are commonly used to measure the quality of graphene transparent electrodes. These two parameters can macroscopically characterize the defects of graphene films transferred from a metal substrate to a quartz transparent substrate,[17, 18] including the doping of graphene films by external factors such as etching solution and the damage caused during the transfer.[19] But because of the conductivity and opacity of the metal substrate, for graphene films grown on a metal substrate, the quality of graphene and graphene films cannot be characterized directly by measuring transmittance and block resistance. However, for graphene films grown by the same process in different laboratories, or graphene films obtained in the same laboratory by different growth processes, there is a lack of reasonable and reliable criteria for comparing coverage, hence other indicators need to be explored to characterize the quality of graphene films. Research shows that the growth conditions of graphene films can be optimized by the coverage of graphene films on the metal substrate, and improving the coverage of graphene films on metal substrates can greatly improve the conductivity and other properties of the transferred graphene films. This section proposes to establish the SEM method for measuring the coverage of graphene films using the characteristic parameters for the coverage of graphene films on metal substrates.

Graphene has high-temperature resistance and oxidation resistance. Thus, when a metal substrate sample grown on graphene film is heated in the atmosphere, a metal substrate not covered by graphene film will form an oxide, while a metal substrate covered by graphene film is protected from oxidation by the graphene film. Hence, the contrast between the covered and uncovered areas of the graphene film can be improved, and according to this characteristic,

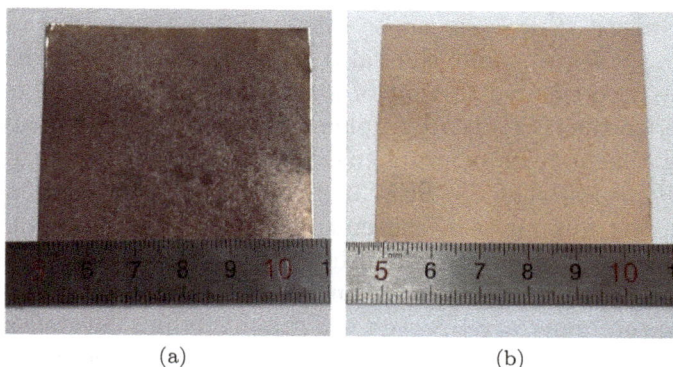

(a)   (b)

Fig. 6.12. Picture of 5 cm × 5 cm copper-based graphene film samples before and after oxidation treatment: (a) the macro-morphology before oxidation treatment; (b) the macro-morphology after oxidation treatment.

the coverage of graphene film can be judged preliminarily. Figure 6.12 is a picture of a sample of 5 cm × 5 cm copper-based graphene before and after oxidation treatment. The oxidation temperature is 180°, and the oxidation time is 5 minutes.[20, 21] When the adjacent graphene film is not covered by an area or the defect size reaches the order of millimeters, the covered and uncovered areas of graphene can be clearly distinguished by the naked eye alone,[22] and the coverage can then be calculated using an optical microscope. When the size of the uncovered area or defect is in the order of microns or even nanometers, a measurement tool of higher resolution must be selected. SEM is an effective measurement technology, but the problem with SEM is that the measurement range is very small. For macroscopic graphene film samples, the ideal coverage result should be expressed as the ratio of the sample area covered with graphene film to the total sample area. In order to reflect the micro-morphology of a 5 cm × 5 cm sample, 8,000 images of SEM at the same multiple are needed. The measurement efficiency is too low, so it is necessary to establish an efficient measurement method.

The following is an introduction to a method for quantitatively characterizing the coverage of graphene films on metal substrates using SEM. In the SEM image, the areas covered and uncovered by graphene films do not have the same contrast, so the pixel area of different contrast degrees can be calculated respectively using image processing software, and the ratio of pixel area of an image of a graphene-covered area to that of an uncovered area can be used to characterize the coverage of graphene films on a metal substrate. By studying the number of graphene domains selected from the samples, it is proved that the SEM method is effective and feasible. By studying the minimum measurement sample size and sample standard deviation, the coverage and homogeneity of the whole thin film sample can be obtained from a limited number of samples. And by studying the boundary between the graphene film covered area and the uncovered area, the uncertainty introduced to the coverage calculation results by the boundary of the graphene film covering area is given. The inspection proves that we can use only a limited number of micro-area coverage to reflect the coverage of the whole film. This method is accurate, fast and convenient.

6.1.3.1   *Measurement principle*

The coverage of graphene film is the ratio of the area of the substrate covered by graphene film to the total area of the substrate. The substrate area covered by graphene film in an SEM image is the product of the number of pixels and the area per pixel in the graphene imaging area. The total area of the substrate is the product of the number of pixels of the whole substrate and the area per pixel, therefore, the coverage of graphene film can be calculated according to Equation (6.14):

$$\theta = \frac{CP}{TP} \times 100\% = \frac{C}{T} \times 100\% \qquad (6.14)$$

where θ is the coverage of the graphene film, C is the number of pixels in the area covered by graphene in the SEM image, T is the number of pixels in the SEM image and P is the area per pixel. It is clear from the equation that for the same SEM image, the area per pixel is the same at all locations, so the area per pixel can be reduced. Therefore, the coverage of graphene film measured by SEM is only related to the number of pixels.

6.1.3.2 *Photographing and image analysis of SEM*

6.1.3.2.1 Sample preparation

Cut a $2\,cm \times 2\,cm$ square sample from a copper-based graphene film sample, and use a conductive adhesive to glue the square sample onto a suitable table for testing. Keep the environment and utensils as clean as possible during sample handling to prevent copper foil from deforming and producing contaminants. For wrinkled samples, lay them flat between two clean slides, then press the slides to flatten the copper foil and cut the sample.

6.1.3.3 *Calibration and measurement of scanning electron microscopy*

First, calibrate the SEM according to Section 6.1.1, then place the graphene sample into it, focusing the SEM and adjusting the brightness and contrast to maximize the difference in gray values between graphene and copper substrates, and finally take a high-resolution SEM image. Select SE mode, Inlens mode, ETD mode or CBS mode as the image mode, and select 3–5 kV as the acceleration voltage. If the graphene coverage is too high, the ETD pattern cannot easily distinguish the graphene coverage from the uncoverage area; the CBS mode sensitive to elements can be selected to filter out interference with the contrast of the covered and uncovered areas due to surface undulation of the substrate.

Select 5–9 different areas of the graphene sample for measurement; each area must contain 15–30 graphene islands or gap islands to be representative of the area. When the coverage is very small ($<3\%$) or very large ($>97\%$), because of the relatively small total area of the graphene island or gap island area, the selected area may contain only five to ten graphene islands or gap islands.

6.1.3.4 *Calculation of coverage*

Use Photoshop software to check each box, and identify the boundary of graphene island in the SEM image. The number of pixels in the boundary is the number of pixels C in the area covered by the graphene film and the number of pixels in the whole image is T.

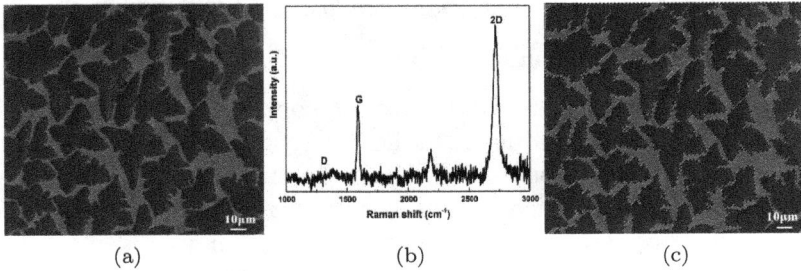

Fig. 6.13.   Graphene film sample on copper foil substrates: (a) SEM image; (b) Raman intensity figure; (c) SEM image selected by Photoshop software from the dark areas (areas covered by graphene films) in (a).

Figure 6.13(a) is the SEM image of graphene film samples on a copper foil substrate in ETD mode. Figure 6.13(b) is a Raman intensity figure of a graphene film sample on a copper foil substrate. There are characteristic peaks G and 2D of graphene in the image, which indicates that the dark areas of the sample are areas covered by graphene films. Figure 6.13(c) is the SEM image after adjusting the tolerance values by Photoshop software and using the magic wand tool to gradually and comprehensively select the dark areas (areas covered by graphene films) in Fig. 6.13(a). Each graphene film covered area is surrounded by a white dotted line, which is defined as an island. It is clear from Fig. 6.13(c) that the selected boundary is clear and fits well with each graphene island. After the coverage area is selected, the number of pixels of the covered area $C = 6,917,382$ and the total number of pixels in the image $T = 8,888,880$ will be displayed in the "Record, measure" part under the software interface. The coverage of graphene film in the SEM image is calculated according to Equation (6.14) as $\theta = 77.8\%$.

It can be seen from the above calculation that the calculation of coverage depends on measuring a limited number of SEM images to count the macroscopic coverage of graphene films. A study is therefore needed to determine how many graphene islands should be present in an SEM image. To ensure that it is reasonable to give the number of graphene islands, the researchers looked at the source and magnitude of the uncertainty to determine the number of graphene islands that should be included in a valid SEM image. The uncertainty was analyzed according to the mathematical model of the coverage calculation and the influence and result of uncertainty sub-item were introduced into the study:

$$C = \overline{S}N \tag{6.15}$$

where S is the average area of the graphene island and N is the number of graphene islands.

Substituting Equation (6.15) for Equation (6.14), we can get

$$\theta = \frac{\overline{S}N}{T} \times 100\% \tag{6.16}$$

Under the measurement circumstance of using the same equipment and at the same magnification, T is a constant. According to the uncertainty transfer equation, we can get

$$\sigma_\theta = \sqrt{\frac{\partial f^2}{\partial \overline{S}}\sigma_{\overline{s}}^2 + \frac{\partial f^2}{\partial N}\sigma_N^2 + \frac{\partial f^2}{\partial T}\sigma_T^2} \tag{6.17}$$

where σ_θ is the uncertainty of graphene coverage, $\sigma_{\overline{s}}$ is the uncertainty introduced for the calculation of the average area of graphene islands, σ_N is the uncertainty introduced into the measurement range due to measurement of the number fluctuation of graphene islands and σ_T is the uncertainty introduced in measuring the total area. Since the image pixel area is a constant when the sample is measured at the same magnification, its uncertainty is $\sigma_T = 0$. According to Equation (6.17), we can get

$$\sigma_\theta = \frac{\sqrt{N^2\sigma_{\overline{s}}^2 + \overline{S}^2\sigma_N^2}}{T} \tag{6.18}$$

It is clear from Equation (6.18) that the smaller the uncertainty caused by the island area and the number of islands, the smaller the effect on the accuracy of the coverage measurement. For a single island of graphene, the introduced uncertainty is its perimeter times a relative width. That is,

$$\sigma_{\overline{s}} = lR \tag{6.19}$$

where l is the perimeter of the graphene island and R is the resolution of SEM. Since l is proportional to the square root of the area of the graphene island,

$$l = \alpha\sqrt{\overline{s}} \tag{6.20}$$

where α is a coefficient related to the shape of the island. The number of graphene islands in the collection area also affects the coverage statistics. Suppose that the number of graphene islands is a random process, then

$$\sigma_N = \frac{1}{\sqrt{N}} \tag{6.21}$$

Substituting Equations (6.19)–(6.21) for Equation (6.18), we can get

$$\sigma_\theta = \sqrt{\frac{(\alpha R)^2}{T}\theta N + \frac{\theta^2}{N^3}} \tag{6.22}$$

It is clear from Equation (6.22) that θ is a function of N, and there is a minimum, that is, taking a certain number of graphene islands within sight will minimize the uncertainty in the calculation of graphene coverage. In order to study the number of inner islands in SEM images with minimum uncertainty, three samples with respective graphene film coverage of 20%, 80% and 90% were conducted through theoretical simulation and experimental measurement. In theoretical simulation, R, T and α use estimated values. Reduce the island to a square whose circumference is four times the square root of the area, estimate α by 4; R^2/T is the resolution of the edge of the sample for the entire image, R is the resolution of a point for an SEM device which is $1/1{,}000$ of its total side length, and then take R^2/T as $1/100{,}0000$. Although these numbers were rough, we found that these estimated values did not have a significant effect on the results after trying other values. It is clear from Fig. 6.14 that the change trend predicted by the theory is basically consistent with the experimental data. A comparison of the theoretical values and experimental data in Fig. 6.14 shows that taking the value of N as 15–30 is rather proper, and the uncertainty calculated from the experimental results is 1–4%.

6.1.3.5 *Number of valid SEM images required to measure the macroscopic graphene film coverage*

In statistics, unless the sample is known to be homogeneous or some analytical questions require representation of the sample, a sufficient number of samples must be analyzed to guarantee the reliability of the results. In order to estimate the minimum sample size, the sample

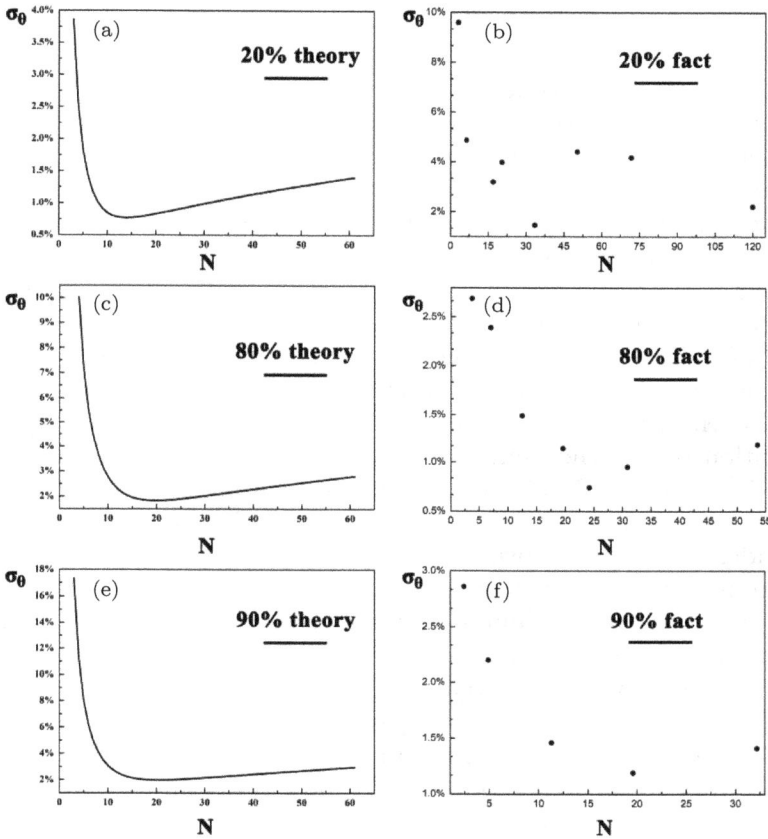

Fig. 6.14. Relationship of σ_θ and N under different coverages: (a) (c) (e) theoretical simulation curve under coverage 20%, 80% and 90%; (b) (d) (f) experimental measurements under coverage 20%, 80% and 90%.

variance is obtained by measuring the sample size, and then the minimum number of samples n required to reach a certain confidence level is calculated according to Equation (6.23):

$$n = \frac{t^2\sigma^2}{Q^2\overline{X}^2} \qquad (6.23)$$

where t is the value at the required confidence level, σ is the population standard deviation of a measurement sample, X is the mean value and Q is an acceptable relative percentage deviation from the mean value. For the confidence level of 95%, we can take the initial t at 1.96 and calculate the value of n by it.

In the process of growing graphene films, there are tiny temperature differences or different gas environments between each basement area, so the coverage of graphene film varies with the substrate area. So, for a sheet of 5 cm × 5 cm graphene that needs to be measured (population), the method of isometric sampling of a nine-palace lattice was adopted. The coverage of graphene films in these nine areas was measured at magnification, and the mean value of the data set $\overline{X}$ and the population standard deviation σ were obtained. Using Equation (6.23), the minimum number of samples n_o required at a certain confidence level is obtained. If the minimum number of samples n_o is less than 9, the average coverage of these samples is then considered to be the coverage of the graphene film (population). If the minimum number of samples n_o is larger than 9, resample and measure based on the minimum number of samples calculated, that is, divide the sample into n_o equal parts, and judge according to the above rules until the requirements are met.

Measure the 5 cm × 5 cm sample in Fig. 6.12, sample it according to nine grid isometric sampling method, the area of each square is (5/3) cm × (5/3) cm, number these squares from A_1 to A_9, and cut another 2 mm × 2 mm square in the middle of each square, as an SEM sample. Number these squares from a_1 to a_9, the sample numbers are shown in Fig. 6.15.

Perform high-resolution scanning imaging of samples a_1–a_9, taking the determination of coverage of sample a_1 as an example. First, observe the whole surface of a_1 to be measured at low magnification, then make a rough estimate of the level not covered, and finally, select an appropriate magnification to image a_1; the best magnification is able to distinguish clearly the boundary between the covered and the uncovered areas and ensure that the SEM image has sufficient number of islands. Image different areas of a_1 multiple times to obtain multiple SEM images of a_1. Use the same magnification to obtain a

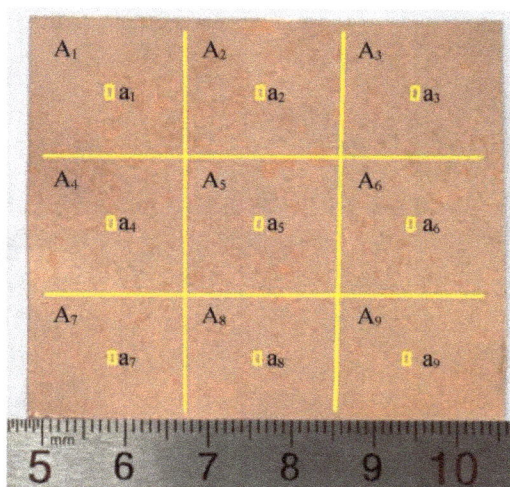

Fig. 6.15. No. of samples to be tested which is selected by the method of isometric nine-grid sampling.

series of a_2–a_9 SEM images. As shown in Fig. 6.16, one SEM image of each of the samples a_1–a_9 is taken at the same magnification.

When using the Photoshop magic wand tool to select an uncovered area, set the tolerance to 10 and get the pixel data. According to Equation (6.14), the coverages of the samples a_1–a_9 are respectively 76.2%, 84.5%, 82.2%, 79.3%, 84.9%, 79.4%, 81.7%, 81.9% and 79.1%. From this, the sample standard deviation σ of coverage of the samples a_1–a_9 is obtained at 2.6%; the mean value of coverage X is 81%. According to Equation (6.23), when the confidence level is 95% ($t = 1.96$), suppose the relative standard deviation assumed to be acceptable for sample homogeneity Q is 6%, and the number of samples required $n = 1.13$. This means that the whole graphene sheet is covered very evenly; in theory, only two SEM images of a particular area of the film need to be measured; the obtained coverage can reflect the coverage of the whole sheet of graphene film. This proves that the mean value obtained from the nine-grid sampling is representative enough; therefore, the coverage of the whole sample film is 81.0%.

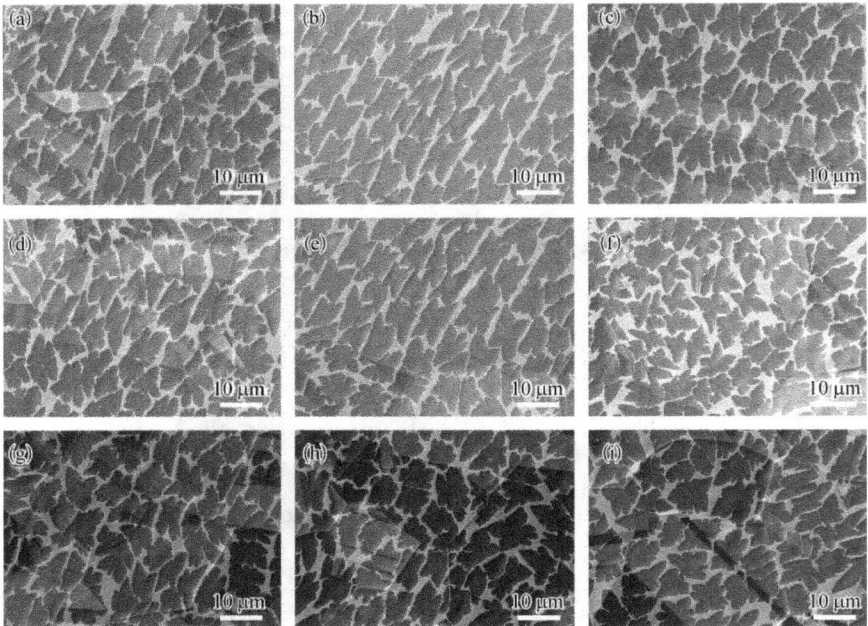

Fig. 6.16.   SEM images of the same magnification obtained respectively from the samples a_1–a_9.

6.1.3.6   *Expression of coverage homogeneity of macroscopical graphene films*

The homogeneity of macroscopical thin films h is expressed by the coverage parameter, see Equation (6.24):

$$h = 1 - \frac{\sigma_u}{\theta} \times 100\% \qquad (6.24)$$

where σ_u is the uncertainty introduced by the inhomogeneity of the sample on the large area substrate; it is calculated according to Equation (6.25). For large-area graphene film samples, in the case of a valid SEM image with 15–30 islands, using the method of isometric nine-grid sampling can meet the representative requirements for homogeneity evaluation of 5 cm × 5 cm samples or even larger samples. For macroscopic samples, the standard deviation obtained by measuring the coverage of a nine-grid sample σ_l consists of two parts: standard deviation for measuring coverage from multiple positions in a nine-grid lattice σ_s and uncertainty introduced due to the inhomogeneity of the sample over long distances σ_u. That is,

$$\sigma_u = \sqrt{\sigma_l^2 - \sigma_s^2} \qquad (6.25)$$

Take, for example, the graphene sample in Fig. 6.17; the homogeneity of a sample with a coverage of about 20% is calculated by using the method of nine-grid sampling. First, measure σ_s in any of the nine grids. Then, measure σ_l by selecting the measuring point in a wide range of samples using a nine-point method to calculate σ_u and the homogeneity of sample coverage h. According to Equation (6.14), the coverage in any one of the nine-grid lattices is obtained as $\theta = 24.7\%$, the standard deviation of coverage $\sigma_1 = 2.0\%$, coverage obtained by the method of nine-grid lattice sampling $\sigma_l = 24.8\%$ and standard deviation $\sigma_l = 2.6\%$; substitute it in Equation (6.25) to get $\sigma_u = 1.6\%$; the homogeneity of sample coverage is calculated to be $h = 93.5\%$.

In summary, the combination of SEM and image processing software is a reliable method to measure coverage. Select the samples of graphene films by the method of nine-grid isometric sampling; it can

Fig. 6.17.   SEM image of sample of about 20% coverage.

be known through theoretical simulation and experimental measurement that when the number of graphene islands in the selected SEM image is 15–30, the uncertainty of graphene film coverage is minimal (1–4%). The coverage of the graphene film in the micro area analyzed statistically using SEM images expresses the process of calculating and sampling the effective number of SEM images necessary for the macroscopic coverage of graphene films. The standard deviation of coverage obtained based on a quantitative comparison between long-range sampling and short-range sampling gives the quantitative expression equation of macro-graphene films covering homogeneity. The coverage and coverage homogeneity measurement method has saved time and ensured the effectiveness of measurement. Moreover, this method can be applied to the image processing of STM and AFM. It can also be applied to the image processing of other two-dimensional materials.

6.2 Measurement of Graphene Morphology, Layer Number and Layer Distance by Transmission Electron Microscope

TEM is one of the most important devices for studing the microstructure, morphology and composition of materials; its physical and structural drawings are shown in Fig. 6.18. The principle of TEM is as follows: Electron gun emits electrons in ultra-high vacuum, and the electrons are accelerated by ten thousand to mega volts. After passing through the electromagnetic lens system, they can converge or illuminate parallel on a sample of nanometer thickness. The electrons are scattered by the electrostatic potential of the atom in the sample when they pass through it, causing discrete electron scattering and continuous electron diffraction to occur to obtain a contrast image.

The imaging mechanism of TEM is as follows: The sample thickness and atomic number are different in different areas of the sample, and the scattering and absorption of electrons are also different, resulting in a mass-thickness contrast. The different area in the crystal sample satisfies the Bragg diffraction condition to a different

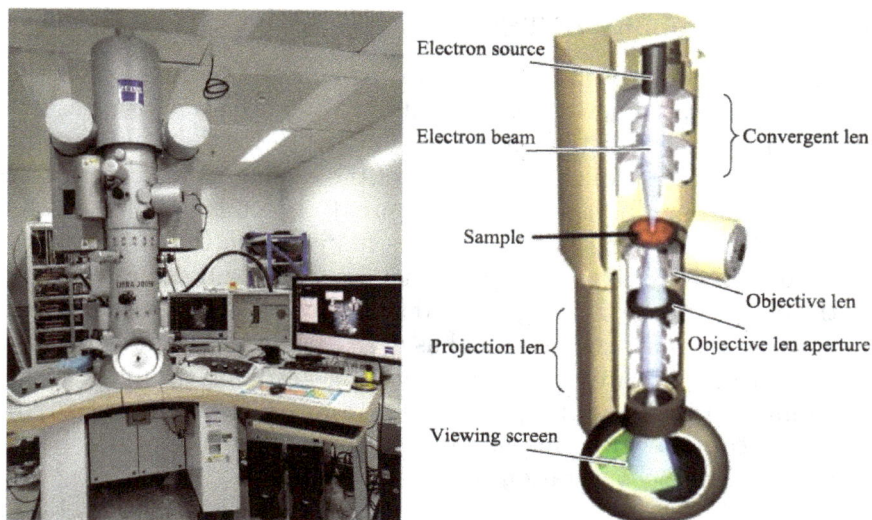

Fig. 6.18.   Physical drawing and structural drawing of TEM.

extent, and the intensity ratio of the diffracted and transmitted beams is different, resulting in a diffracted contrast. When a parallel electron beam penetrates a crystal sample, the phase of the transmitted and diffracted electron beams is modulated by the periodic crystal potential field so that the outgoing electron beam carries the structural information of the crystal to produce phase contrast. It collects highly angular incoherent scattered electrons from the atomic column in the sample to eliminate the relevant information. The image point intensity of the atomic column in the image is proportional to the square of the average atomic number of the atomic column to produce atomic number contrast.

TEM has extremely high resolution, and it can obtain simultaneously all the information about the morphology, chemical composition, crystallography, microstructure, etc., thus enjoying wide and important application in the field of material research. TEM with various analytical functions is particularly favored by researchers. High-resolution imaging and high-angle annular dark-field imaging can measure and study the morphology and structure of materials at the atomic scale.

6.2.1 Traceability of TEM

TEM can accurately measure the microstructure at nanoscale; the characteristic parameter closely related to the accuracy of measurement is magnification. The magnification of TEM is the scaling factor of image magnification; the accuracy of the scaling factor determines the accuracy of the size of the measured object in the image. As TEM is influenced by factors such as filament lifetime, accelerated voltage, current stability, magnetic field stability and sample quality in the process of use, its magnification will deviate with the changes in working time. At present, China has over 1,000 TEMs and is increasing at an annual rate of over 100, moving gradually from scientific experimentation to actual production. In order to ensure the accuracy and consistency of measurement results in the social R&D and industry, periodic calibration of TEM is required.

China has formulated some standards and specifications for TEM in the 1990s, such as JB/T 5584-1991 *Methods of Measurement of TEM Magnification* and JY/T011-1996 *General Principles of TEM*. These standards and specifications provide some basic definitions of terms and general principles of measurement methods. In 1997, the National Education Commission issued JJJG011-1996 *Verification Regulation of TEM*; this regulation is applicable to the verification of all kinds of TEM after new installation, use and maintenance. The verification items stipulated by this regulation for TEM include lattice resolution, point resolution, magnification, image distortion, contamination rate, drift rate and vacuum degree. But this regulation was issued and implemented more than two decades ago when China had not issued relevant CRMs. As a result, the verification regulation of TEM cannot be implemented effectively. With the advancement of science and technology and the rapid development of electron microscopy, in 2017, General Administration of Quality Supervision, Inspection and Quarantine and China National Standardization Administration issued national standard GB/T 34002-2017/ISO 29301:2010 *Image Magnification Method for Calibration of Periodic CRMs for Microbeam Analysis in TEM*. The standard only specifies the most important parameters for the calibration of the metrological characteristics of TEM — calibration of magnification — but it is not applicable to dedicated critical size measurements of length

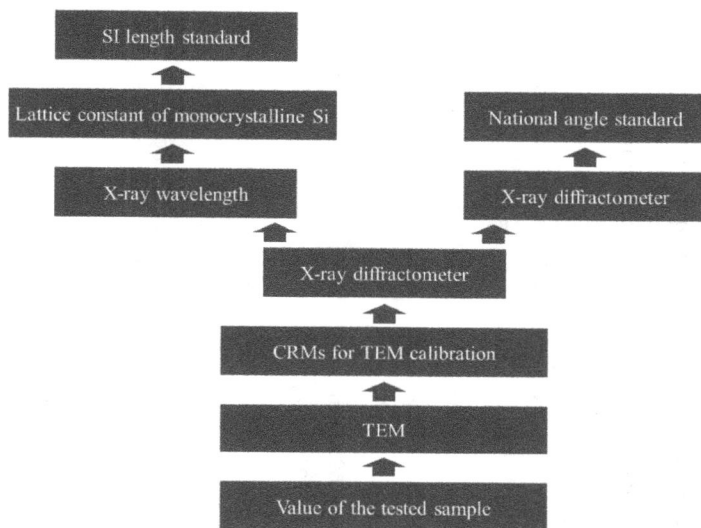

Fig. 6.19.   Traceability of TEM established by National Institute of Metrology, China.

by TEM and scanning TEM; there is no calibration for characteristic parameters of image distortion, pollution rate and drift rate. Figure 6.19 is the tracing path of TEM magnification established by National Institute of Metrology, China. It is clear from the figure that the certified reference material used for magnification calibration adopts an X-ray diffractometer traceable to the SI basic unit to set the value, see Chapter 4 for the calibration of X-ray diffractometers. National Institute of Metrology, China, has developed and published the reference material of gold film crystal plane distance (GBW13655), the only certified reference material for high magnification calibration of TEM at present. Other research institutions in China have developed and issued CRMs for medium and low magnification calibration, see Table 6.7. CRMs are carriers of accurate value transmission; through CRMS, the characteristic parameters of TEM are calibrated according to the calibration standard, the uncertainty of calibration result is evaluated, and the exact value traced to the SI basic unit is passed

Table 6.7. Certified reference materials for TEM magnification calibration developed in China.

Name	Unit of development	Magnitude characteristics	Measurement range
Reference material for crystal plane distance of gold film (GBW13655)	National Institute of Metrology, China	Crystal plane distance 0.2366 nm; Uncertainty 0.0030 nm	High magnification
Reference material for Silicon dioxide film thickness (GBW13965)		Film thickness 9.92 nm Uncertainty 0.40 nm	Middle magnification
Reference material for thickness of superlattice multilayer films (6–10 layers, GBW13955)		Film thickness 10.60 nm Uncertainty 0.18 nm	Middle magnification
Reference material for thickness of silicon nitride thin films (GBW13961)		Film thickness 52.67 nm Uncertainty 0.28 nm	Low magnification
Reference material of gold particle size [GBW(E)120 126]	National Center for Nanoscience	Particle diameter 22.6 nm	Middle magnification
Reference material of gold nanorod particle size [GBW(E)140474]		Uncertainty 1.0 nm	Low magnification
Reference material for particle size of polystyrene (GBW12011)	China University of Petroleum	Particle diameter 79.1 nm Relative uncertainty 1.9%	Low magnification

Table 6.8.   Main metrological characteristics and reference material requirements for TEM calibration.

Measurement characteristics	Magnification (M)	Expanded uncertainty of CRMs ($k = 2$)
Magnification indication error	$M > 300{,}000$	Less than 1.5%
	$100{,}000 < M \leq 300{,}000$	Less than 2%
	$M \leq 100{,}000$	Less than 2.5%
Sample contamination rate	$M \leq 100{,}000$	Less than 2%
Sample drift rate	$M \leq 100{,}000$	Less than 2%

to the end user. This process is described in detail in Section 6.22. Based on the study of calibration and tracing of TEM, National Institute of Metrology, China, has developed the group standard of Chinese Society for Testing and Materials (CSTM), Zhongguancun, T/CSTM00162-2020 *Calibration Method of TEM*. In addition to the specification of metrological characteristic parameters for TEM calibration, the evaluation method of uncertainty is also given. The following is an introduction to the calibration method. The main metrological characteristics and reference material requirements for TEM calibration are shown in Table 6.8.

6.2.2   Certified Reference Materials for the Magnification Calibration of TEM

The magnification of TEM is continuously adjustable from a few thousand to one million times and is usually divided into three levels: low, middle and high (less than 100,000 times, 100,000 to 300,000 times, and more than or equal to 300,000 times, respectively). Therefore, three kinds of CRMs are needed to calibrate the low, middle and high magnification of TEM. Callibration of TEM magnification requires CRMs. China has developed a reference material for TEM magnification calibration, and the national resource sharing platform for CRMs (https://www.ncrm.org.cn) provides information on CRMs available in China. The CRMs for TEM magnification calibration mainly include crystal plane distance, film thickness and particle size CRMs. Table 6.7 lists the name and range of value of relevant rectified reference materials in China. When the magnification of TEM is calibrated, the reference material corresponding to the reference value can be selected for calibration according to the range of the values to be measured.

There is very little research on the CRMs for TEM magnification calibration in foreign metrology institutes. US NIST has developed several CRMs for gold particles: for gold sol (SRM 8011), the reference value for particle size is 10 nm; for gold sol (SRM 8012), the reference value for particle size is 30 nm; for gold sol (SRM 8013), the reference value for particle size is 60 nm. They are mainly used for middle and low magnification calibration in TEM, and there is no reference material for high magnification calibration of TEM yet. In addition, there are some TEM high magnification calibration samples on the market at home and abroad. However, these calibration samples do not have a traceability description nor are they certified as CRMs, as is shown in Table 6.9.

It is very important to select and use the CRMs for TEM high magnification calibration. Some TEM users use the crystal plane distance of gold particles or other crystal samples to calibrate high magnification of TEM; the crystal plane distance values of these samples are not traceable or accurate. The author/researcher found that during the growth or preparation of particles or films, the strain of different crystal orientations is different, which leads to the difference between the measured value and the theoretical value of crystal plane distance. As is shown in Fig. 6.20, the crystal lattice distortion

Table 6.9.   Foreign calibration samples of high magnification for TEM.

Calibration sample	Development institution	Measurement characteristics	Measurement range
Si/SiGe	NRC Institute for Microstructural Sciences (NRC-IMS)	The crystal plane distance is 0.31356 nm Film thickness is 9.2 nm	High, middle and low magnification
Single Crystal Gold	Tedpella	The crystal plane distance is 0.23548 nm	High magnification
Catalase		The crystal plane distance is 0.9 nm	High magnification

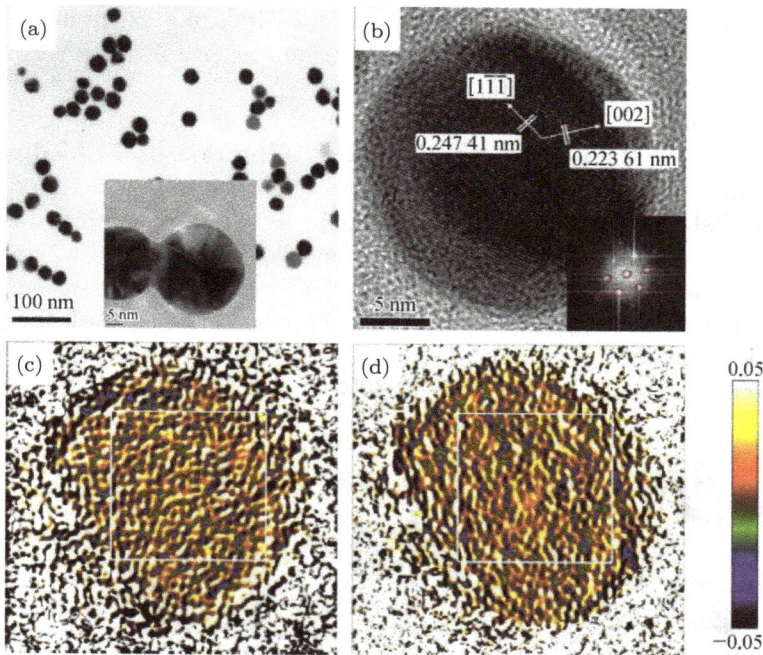

Fig. 6.20. Microstructure and strain distribution of gold particles: (a) bright field image; (b) high-resolution crystal lattice image; (c) strain distribution of {111}; (d) strain distribution of {002} plane strain distribution.

in the process of gold particle synthesis leads to large crystal plane strain; Au{002} crystal plane strain reaches 3.7%. In addition, the crystal plane strain state is unstable and changes with the electron microscope acceleration voltage and irradiation time. Therefore, the crystal plane strain of gold particles is large and the strain state is unstable; it is not suitable to develop high magnification calibration reference material for TEM. Some users use the Si{220} crystal plane distance to calibrate the high magnification of TEM because XRD can be used for direct calibration of silicon wafers. But the problem is that after the silicon wafer is calibrated, it needs ion thinning and other processing methods to get the sample suitable for TEM measurement. Yet the processing may cause a change in the strain state of the crystal plane and the introduction of contaminants; these factors may lead to the change of the Si{220} crystal plane distance value. Therefore, XRD can be used for assigning the default value of the Si{220} crystal plane distance, but it is not reliable to calibrate the TEM magnification with the sample processed after being assigned the default value.

National Institute of Metrology, China, has developed the first reference material for high magnification calibration of TEM directly traceable to the SI length definition — reference material for crystal plane distance of gold film (GBW13655) — thereby solving the calibration problem of high magnification of TEM. Figure 6.21 is the microstructure and strain distribution of gold film reference material. It is clear from the figure that the microstructure of the gold film is composed of equiaxed grains, the strain on the crystal plane Au{002} is large and reaches −2.32%, but the strain on the crystal

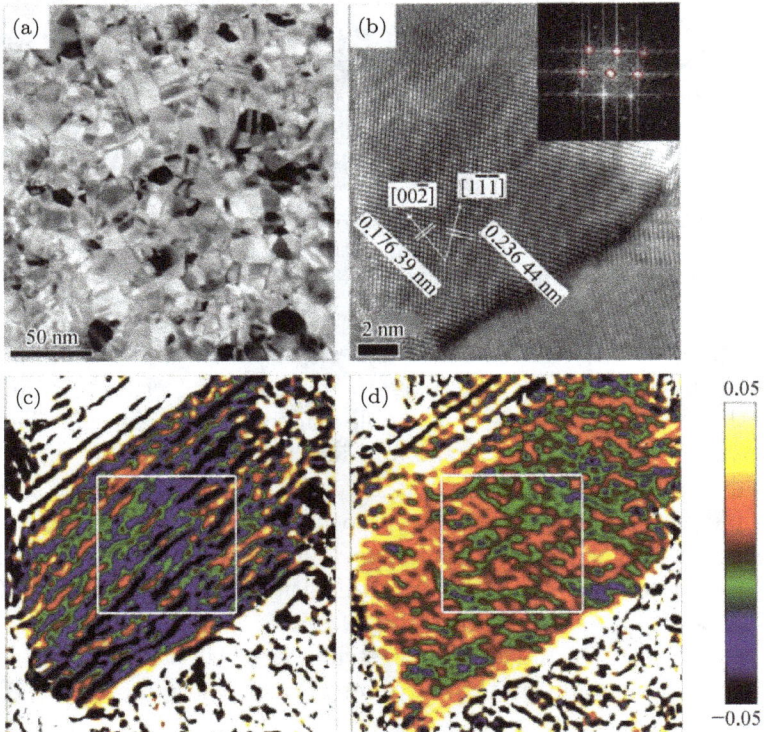

Fig. 6.21.   Microstructure and strain distribution of gold film CRMs: (a) bright field image; (b) high-resolution crystal lattice image; (c) strain distribution of {002}; (d) strain distribution of {111}.

plane Au{111} is small and only reaches 0.26%. More importantly, the gold film is grown directly on the copper mesh carbon film; it can be used directly without any processing after being assigned a default value by XRD, to align the calibration value with the assigned value.

6.2.3 Calibration Methods of TEM

6.2.3.1 *Calibration of magnification*

Magnification is a representative parameter of the accuracy of TEM image indication. Therefore, it is necessary to calibrate the magnification. TEM magnification M is divided into three levels: high ($M > 30,000$ times), middle ($10,000$ times $< M \leq 10,000$ times) and low ($M \leq 10,000$ times). The equations to calculate the measured value of length characteristics of CRMs S, magnification indication error Δ and calibration factor η by TEM are as follows:

$$\Delta = \frac{S - S_0}{S_0} \tag{6.26}$$

$$\eta = \frac{S_0}{S} \tag{6.27}$$

where S is the measurement of the length characteristic quantity of the reference material, S_0 is the reference value of the length characteristic quantity of the reference material and η is the magnification calibration factor.

The high magnification of TEM should be calibrated by crystal plane distance reference material. Take high-resolution lattice images of the reference material with interplanar distance, use the image analysis software (Digital Micrograph, ImageJ, etc.) of TEM to obtain the measured value of the crystal plane distance d (Fig. 6.22), and calculate according to the reference value of crystal plane distance d_o the indication error Δ and calibration coefficient η of high magnification.

The medium and low magnification of TEM should be calibrated by reference material of film thickness or particle size. Take the bright field image of the film thickness reference material, use the image analysis software (Digital Micrograph, ImageJ, etc.) of TEM to obtain the measured value of the film thickness T (Fig. 6.23), and

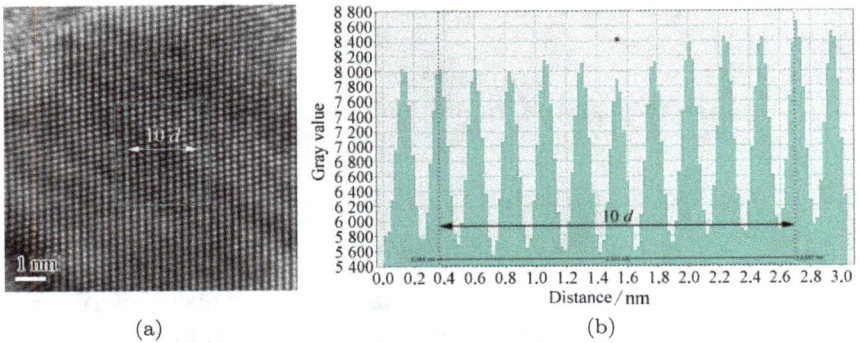

Fig. 6.22.   Measurement of crystal plane distance by TEM: (a) two-dimensional lattice image; (b) gray value spectrum in the vertical direction of {111} crystal plane.

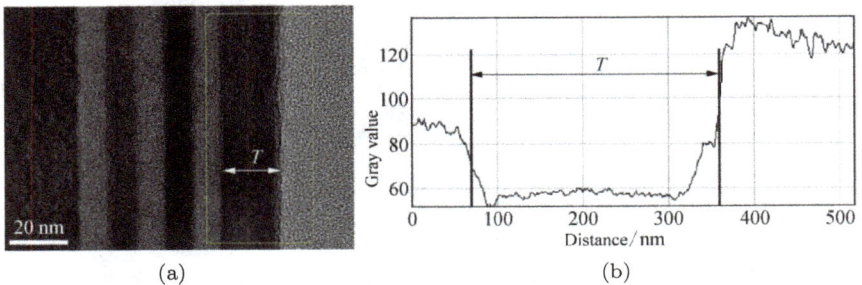

Fig. 6.23.   Measurement of film thickness by TEM: (a) bright field Image; (b) gray value spectrum of thickness direction of thin film.

calculate according to the reference value of film thickness T_o the indication error Δ and calibration coefficient η of high magnification.

6.2.3.2   *Calibration of contamination rate and drift rate*

The contamination rate and drift rate of TEM samples are usually calibrated with the reference material of round hole carbon film. Set the magnification of TEM at 40,000–50,000 times in bright field image mode, move the hole to the center of the screen, adjust the brightness of the image, focus to slightly under focus (refer to the reference value given by the instrument manufacturer), take the first image, keep the parameters of the instrument unchanged, and take

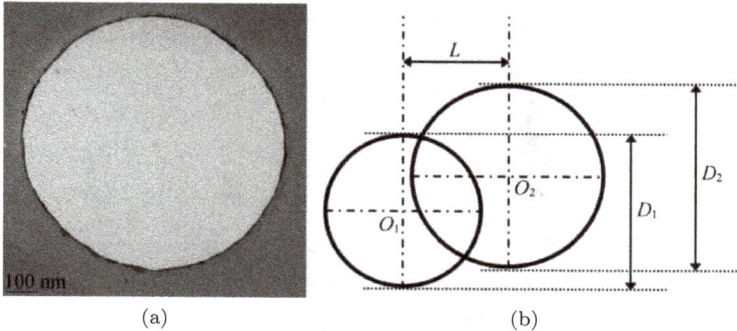

(a)   (b)

Fig. 6.24.   Schematic figure of TEM bright field image of carbon film reference material with round hole and measurement of sample contamination rate and sample drift rate.

the second image 10 minutes later, as shown in Fig. 6.24. Calculate the sample contamination rate and drift rate according to Equations (6.28) and (6.29):

$$Q_c = \frac{D_2 - D_1}{2t} \tag{6.28}$$

$$Q_d = \frac{L}{t} \tag{6.29}$$

where Q_c is the sample contamination rate of the TEM (nm/min), Q_d is the sample drift rate of the TEM (nm/min), t is the time interval between the two images, which is set at 10 min, D_1 is the diameter of the round hole from the first shot, D_2 is the diameter of the round hole from the second shot and L is the distance of D_1 and D_2 from the center of the hole.

6.2.3.3   *Definition of image boundary*

In the process of defining the boundary of SiGe thin films, first, obtain by the Digital Micrograph software the gray-level contour map and original data of SiGe thin film boundary area, and then use the MATLAB software for nonlinear fitting of the original data. The gray-level contour map and gray-level nonlinear fitting map of the right boundary area of SiGe thin films are shown in Figs. 6.25(b) and 6.25(c); its nonlinear fitting function is Equation (6.30); the gray-level contour map and nonlinear fitting map of the left boundary area of

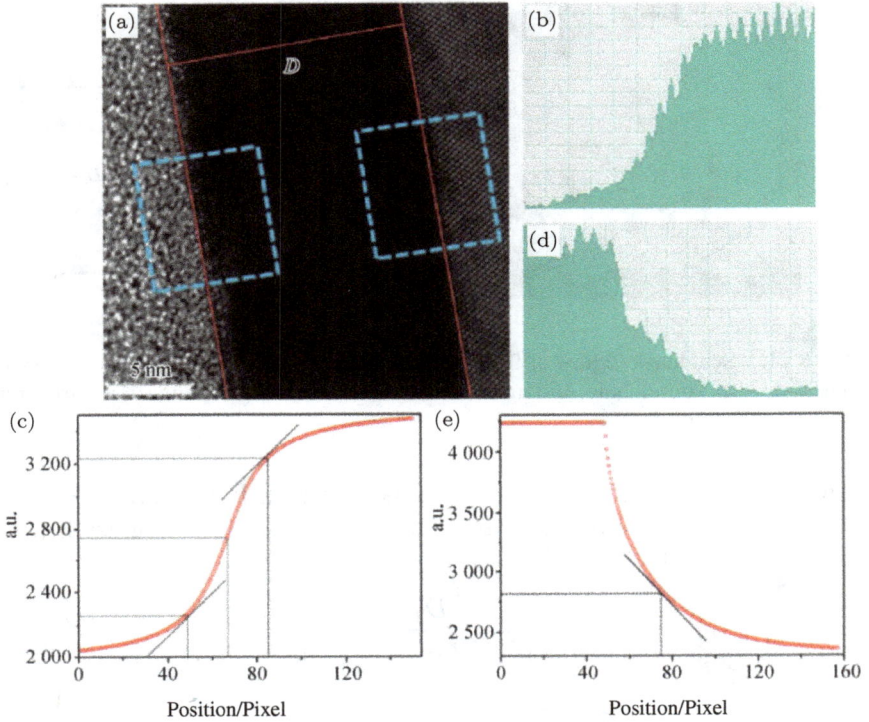

Fig. 6.25.   Boundary delimitation of SiGe thin films: (a) high-resolution lattice image of SiGe thin films; (b) gray line contour map of right boundary region; (c) gray-scale nonlinear fitting graph of right boundary region; (d) gray line contour map of left boundary region; (e) gray-scale nonlinear fitting graph of left boundary region.

SiGe thin films are shown in Figs. 6.25(d) and 6.25(e); its nonlinear fitting function is Equation (6.31):

$$y = a + \frac{b}{\pi}\left[\arctan\frac{x-c}{d} + \frac{\pi}{2}\right] \tag{6.30}$$

where $a = 1{,}942.5333$, $b = 1{,}161.306\ 1$, $c = 66.691119$, $d = 13.226761$ and $r^2 = 0.99$.

$$y = a + b(1 - \exp\{-[x - d\ln(1 - 2^{1/e}) - c]/d\})^e \tag{6.31}$$

where $a = 4{,}233.648$, $b = 1{,}900.75$, $c = 59.524376$, $d = 29.345151$, $e = 0.58673996$ and $r^2 = 0.98$.

The boundary of thin film is a transition area of element diffusion and an area of gradient change of gray value. The boundary is defined as follows: Derive the derivative of nonlinear fitting function, calculate the position where the derivative value equals ±1, and then define the boundary based on the derivative position (or the intermediate position between the two derivative positions), as is shown in Figs. 6.25(c) and 6.25(e).

6.2.3.4 *Uncertainty evaluation of calibration results*

Measurement uncertainty of calibration results for characteristic values such as TEM magnification, contamination rate and drift rate is influenced by noise, magnetic field, vibration, voltage variation, magnetic lens astigmatism and other factors. The uncertainty evaluation of calibration results is as follows:

(1) **A-type uncertainty:** It is mainly the uncertainty u_1 introduced by the repeatability of reference material measurement. Measure the characteristic value at least six times at a certain position of the reference material; the equation for calculating the uncertainty introduced by measurement repeatability is

$$u_1 = \sqrt{\frac{1}{n-1} \sum_{j=1}^{n} (x_j - \bar{x})^2} \qquad (6.32)$$

where n is the total number of repeated measurements, j is the serial number of measurements, $j = 1, 2, \ldots, n$, x_j is the jth measurement and X is the mean value of n measurements.

(2) **B-type uncertainty:** The uncertainty introduced by CRMs for TEM calibration is u_2:

The uncertainty introduced by the spatial resolution of TEM is u_3.

The electron beam drift of TEM is Q, the image exposure time for SiGe film imaging is t, and then the uncertainty caused by the drift of the electron beam u_d is

$$u_d = Q \times t/\sqrt{3} \qquad (6.33)$$

SiGe thin films are epitaxially grown on monocrystalline silicon, and the surface geometry of monocrystalline silicon is tilted.

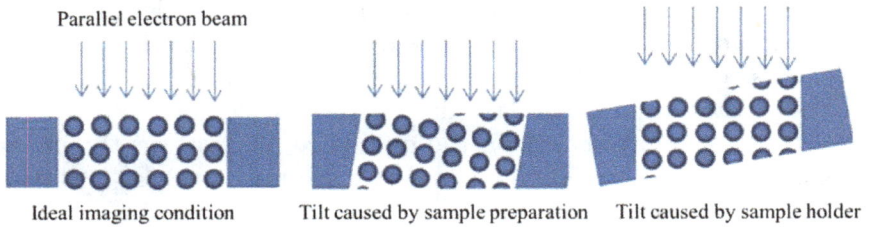

Fig. 6.26.   Schematic figure of tilt correction of TEM samples.

The grinding and polishing of the silicon-germanium film in the preparation process which causes the obliquity are all corrected by the tilt angles of α and β in the sample rod during the measurement, as shown in Fig. 6.26. The uncertainty introduced by sample tilt u_t is

$$u_t = \sqrt{[T(1 - \cos\alpha)]^2 + [T(1 - \cos\beta)]^2} \qquad (6.34)$$

where α and β are the tilt angles of the sample rod and T is the thickness of the SiGe film.

The measured data of TEM are two-dimensional images; the pixel size is determined by both the magnification and the CCD camera's pixels. The pixel size of a high-resolution lattice image of SiGe thin film is P and the uncertainty introduced by pixel size u_p is

$$u_p = P/\sqrt{3} \qquad (6.35)$$

(3) **Combined uncertainty of calibration results:** The components of uncertainty are not correlated; the combined uncertainty is calculated according to the square and root forms:

$$u_{\text{cal}} = \sqrt{u_1^2 + u_2^2 + u_3^2 + u_d^2 + u_t^2 + u_p^2} \qquad (6.36)$$

(4) **Expanded uncertainty of calibration results:** For the normal distribution, when the confidence level is 95%, the

corresponding $k = 2$, then the Expanded uncertainty U is

$$U = k \times u_{\text{cal}} \qquad (6.37)$$

6.2.4 Measurements of Graphene Morphology, Number of Layers and Lattice Spacing by TEM

6.2.4.1 *Sample preparation*

The sample preparation process of thin film includes dissolving the substrate, cleaning the film, scooping up micro-grid carbon film and drying the sample. The sample preparation process of powder measurement includes dispersion, dilution, secondary dispersion, ultra-thin carbon film sample preparation and sample drying. During the preparation of thin film samples and powder samples, the environment and utensils should be kept clean to avoid contamination. The following case describes the preparation of graphene film samples and powder samples.

6.2.4.2 *Preparation of thin film samples*

(1) **Dissolving the substrate:** Cut the warping edge of the graphene film on a copper substrate (about $1\,\text{cm} \times 1\,\text{cm}$), use a sharp pair of tweezers, take a corner, place it in a petri dish containing $20\,\text{ml}$ of $FeCl_3/H_2O$ solution of $1\,\text{mol/L}$ (or $20\,\text{ml}$ of N_2S_8/H_2O solution of $0.2\,\text{ml/l}$), put the copper substrate face down and let the solution slowly corrode the copper substrate.

(2) **Cleaning the film:** After the copper substrate is thoroughly corroded, gently lift the graphene film out of a clean quartz sheet and wash it in anhydrous ethanol repeatedly three times.

(3) **Micro-grid carbon film:** Scoop up the film and put the cleaned graphene film into distilled water, let it float on the surface of distilled water and then scoop up the graphene film with 200 mesh micro-grid carbon film. In the scooping-up process, locate as much as possible the naturally formed crack area on the graphene film in the middle of the micro-grid carbon film, as shown in Fig. 6.27.

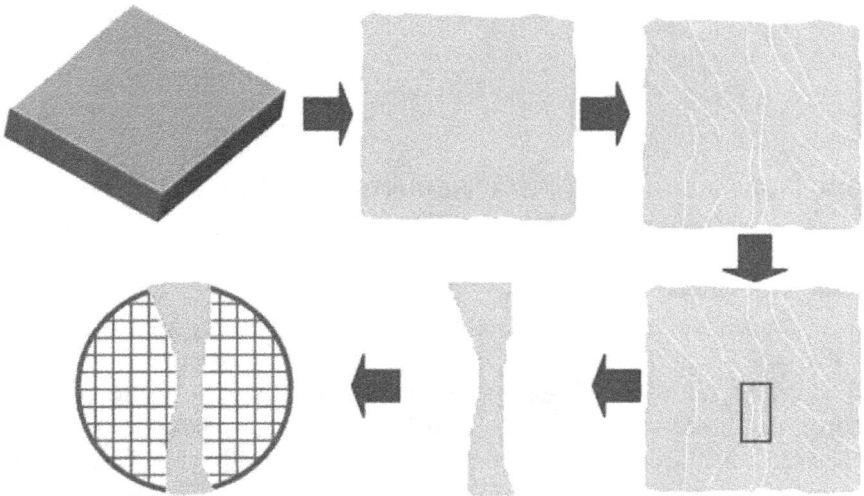

Fig. 6.27. Schematic figure of sample preparation process for TEM of graphene film.

(4) **Sample drying:** After the sample is picked up, place it into the electric heating drum drying box of 45°C to dry 0.5 h and then use it for TEM measurements.

6.2.4.3 *Powder sample preparation*

(1) **Dispersion and dilution:** Use the analytical balance to take 0.002 g graphene powder, pour the powder into a 15 ml centrifuge tube and then fill the tube with anhydrous ethanol. Cover the centrifuge tube and oscillate with the vortex oscillator for 10 minutes to obtain the initial dispersion.

(2) **Secondary dispersion:** Use a dropper to take 1 ml dispersible liquid from the initial dispersible liquid in Step (1) and then fill it with anhydrous ethanol. Cover it with a tube cap, put it on a vortex oscillator and oscillate for 10 minutes.

(3) **Preparation of ultra-thin carbon film sample:** Put a 2,000-mesh ultra-thin carbon film into the centrifuge tube after oscillation of Step (2), keep the ultra-thin carbon film at the bottom of the centrifuge tube and allow the graphene material in the dispersion to settle naturally for 1 h.

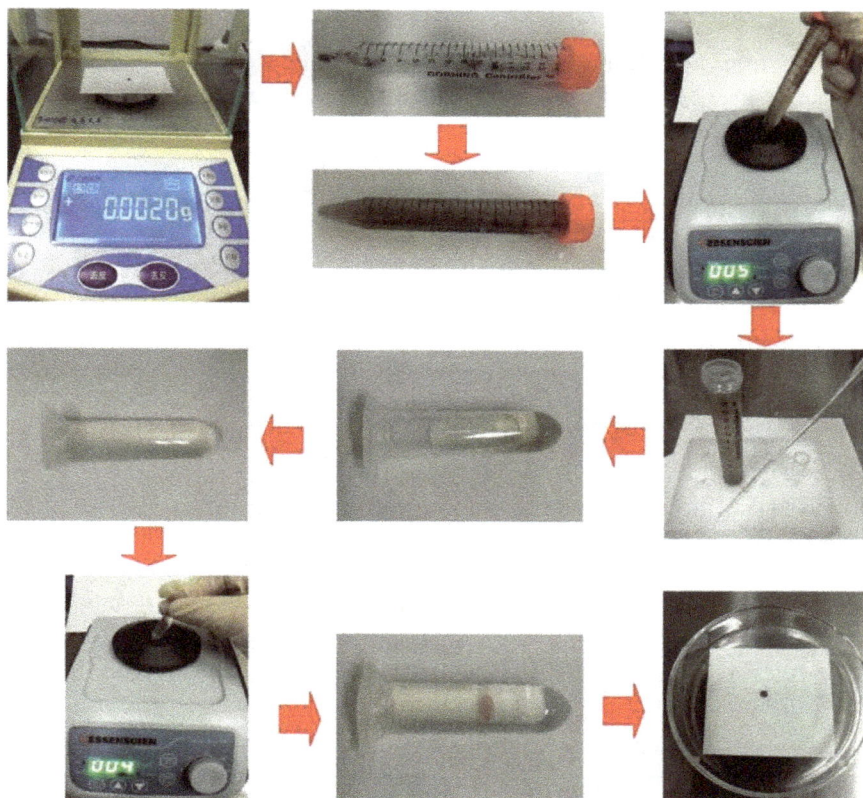

Fig. 6.28. TEM sample preparation of graphene powder.

(4) **Sample drying:** After settling, take out the ultra-thin carbon film and put it into the petri dish, put it into the electric-heated air-blast drying box of 45°C to dry for 0.5 hour and then use it for TEM measurement.

The whole process from Step (1) to Step (4) is shown in Fig. 6.28.

6.2.5 Sample Measurement

TEM shall be calibrated in accordance with T/CSTM 00162-2020 *Calibration Method of TEM* before use.

6.2.5.1 *Selection of measurement position: Film samples*

At low magnification, move the x and y shafts of the sample rod, find the graphene film samples on the micro-grid carbon film and secure the measurement position along the edge of the graphene sample. First, use z shaft for mechanical focusing, then use a magnetic lens magnetic field for focusing, observe and select the measurement position under the condition of slightly under-focus (the reference value given by the instrument manufacturer) selected, as shown in Fig. 6.29(a). Select at least 12 positions to cover as much of the edge of the graphene sample as possible.

6.2.5.2 *Selection of measurement position:*
Powder samples

Divide the whole micro-grid carbon film into four zones. At low magnification, move x shaft and y shaft of the sample rod through four areas in turn. First, use z shaft for mechanical focusing, then use a magnetic lens magnetic field for focusing, observe and select the measurement position under the condition of slightly under-focus (the reference value given by the instrument manufacturer) selected,

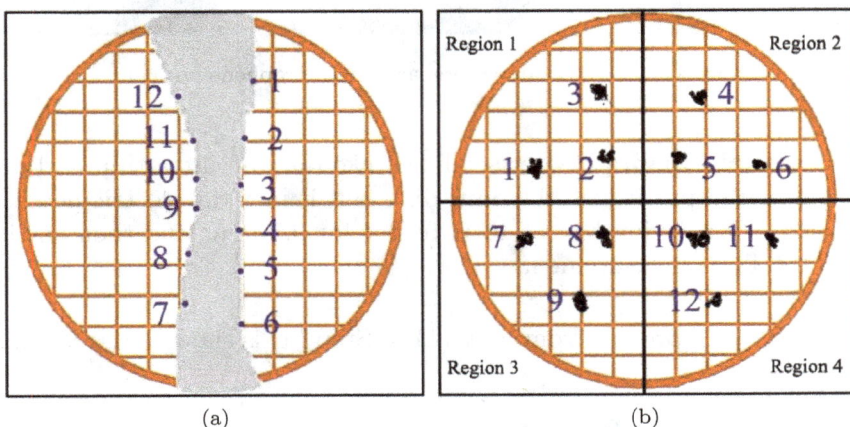

Fig. 6.29. Selection of measurement position for graphene-related materials: (a) graphene film; (b) graphene powder.

as shown in Fig. 6.29(b). Select at least three measurement positions with graphene-related materials for each zone, and the distance between any two measurement positions should be greater than 0.2 mm.

6.2.5.3 *Taking bright field images*

Switch the bright field image mode to the selected area electron diffraction mode at the selected measurement position, select the appropriate objective diaphragm to trap the transmission beam, then switch to bright field image mode, and adjust lighting brightness and focus so that the image brightness is moderate and the contrast best without overexposing the screen and camera. Choose the appropriate magnification, so that the area under test accounts for about two-thirds of the image and set the image pixel to the camera's maximum pixel and the exposure time to 0.5–1s. Take bright field images at different locations and store the images. Bright field images are mainly used to analyze the micrograph and measure the characteristic size of samples and to mark and display the measuring position. Figure 6.30 is the TEM image of graphene powder. It is clear from

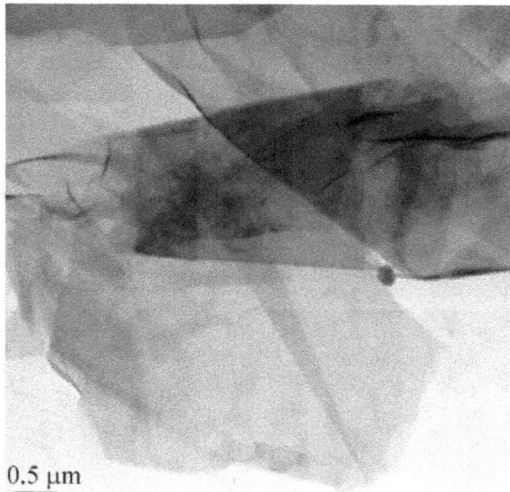

0.5 μm

Fig. 6.30.   TEM bright field image of graphene powder.

the figure that there are crimp, fold and stack morphology in the micro-morphology of graphene samples.

6.2.5.4   *Taking high-resolution lattice images*

Pull out all the stops under bright field image mode, increase the magnification to about 500,000, adjust the inclination of the electron beam, the illumination brightness and focus, set the image pixel to the camera's maximum pixel, and select the exposure time for 0.1–05s. Take high-resolution lattice images of different locations and store the images; all images are required to have the same magnification, the same pixels and the same exposure time. Use a high-resolution lattice image mode of TEM to measure the number and distance of graphene layers. High-resolution lattice images obtained from graphene film comparison experiments (magnification is 600,000, image is 4,008 pixels × 2,824 pixels, exposure time is 0.5s) are shown in Fig. 6.31.

6.2.6   Data Analysis

6.2.6.1   *Layer count*

In high-resolution lattice images of graphene, the lattice fringes are parallel lines of alternating dark and light lines created when

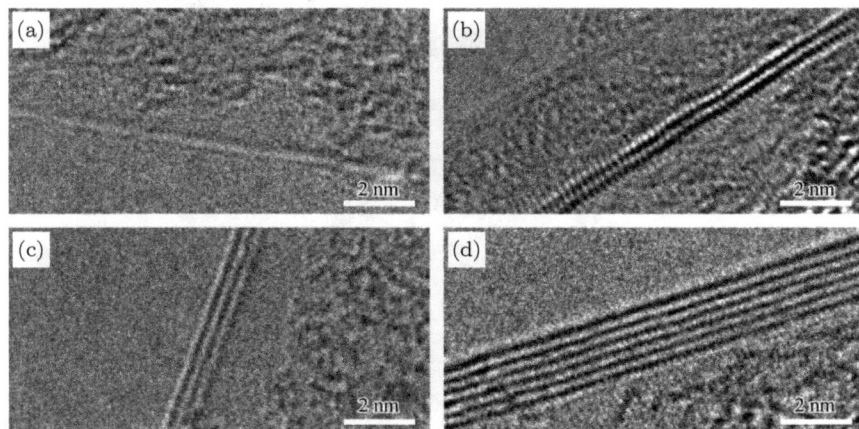

Fig. 6.31.   High-resolution lattice images of graphene films: (a) 1 layer; (b) 2 layers; (c) 3 layers; (d) 6 layers.

electrons are diffracted by graphene layers. A dark line corresponds to a layer of graphene, and the number of dark lines in the lattice fringes is counted by human vision, which is the number of layers of graphene L, as shown in Fig. 6.31.

6.2.6.2 *Lattice spacing count*

Choose the crystal lattice fringe area with good parallelism and clear contrast from the high-resolution lattice image. Use ImageJ or Digital Micrograph software to generate the gray value spectrum of the vertical direction of the lattice fringe and select from the spectrum the intensity peaks between the two edge intensity peaks; the human visual count is N. Measure the overall width W of N intensity peaks ($N - 1$ graphene lattice spacing) using the dimension measurement function of the above software and calculate the average distance between adjacent intensity peaks, that is, the average distance of the graphene layers d_g is shown in Equation (6.38). High-resolution lattice images can be easily obtained because of the folded or curled edges of the graphene material. High-resolution lattice images obtained from graphene film comparison experiments (magnification is 600,000, image is 4,008 pixels $\times$ 2,824 pixels and exposure time is 0.5s) are shown in Fig. 6.32(a). Select the lattice fringe area from the high-resolution lattice image and generate gray value spectrum by ImageJ software (Fig. 6.32(b)). Select from the spectrum all the intensity peaks between the two edge intensity peaks, count $N = 4$, obtain the total width measured according to the above

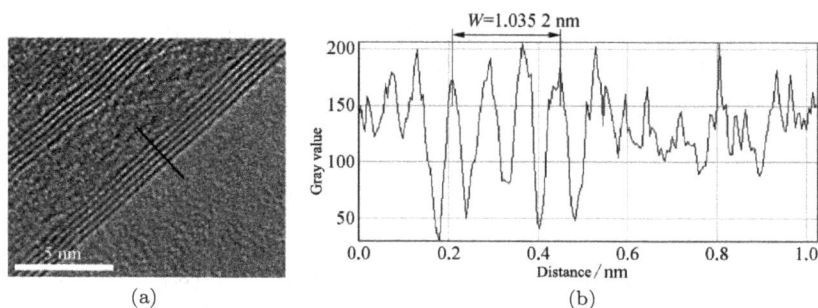

(a)   (b)

Fig. 6.32. Measurement of lattice spacing of graphene films: (a) high-resolution lattice image; (b) the spectrum of gray value in the vertical direction of lattice fringe.

method $W = 1.035\ 2\,\text{nm}$ and calculate the average lattice spacing $dg = 0.3451\,\text{nm}$ according to the Equation (6.38):

$$d_g = W/(N - 1) \qquad (6.38)$$

where d_g is the average distance of the graphene layers (nm) and W is the total width of N intensity peaks (distance of $N - 1$ graphene layers) (nm).

6.2.7   Uncertainty Evaluation

A-type uncertainty

Select at least six positions for measurement of each sample; the uncertainty introduced by the homogeneity of the sample u_1 is

$$u_1 = \sqrt{\frac{1}{m-1} \sum_{i=1}^{m} (x_i - \bar{x})^2} \qquad (6.39)$$

where m is the total number of positions measured, i is the location number, $I = 1, 2, \ldots, m$, x_i is the measurement of ith position and $\bar{x}$ is the mean value of m measurements.

Measure repeatability at least six times at a location in the sample; the standard uncertainty introduced by measurement repeatability u_2 is

$$u_2 = \sqrt{\frac{1}{n-1} \sum_{j=1}^{n} (x_j - \bar{x})^2} \qquad (6.40)$$

where n is the total number of repeated measurements, j is the measurement number, $j = 1, 2, \ldots, n$, x_j is the value of jth measurement and $\bar{x}$ is the mean value of n measurements.

B-type uncertainty

The uncertainty component of TEM calibration is u_{cal}.

Combined uncertainty

The components of each uncertainty are not correlated, and the composite uncertainty u_c calculated by the square and root forms is

$$u_c = \sqrt{u_1^2 + u_2^2 + u_{\text{cal}}^2} \qquad (6.41)$$

Expanded uncertainty

For the normal distribution, when the confidence level is 95% and the corresponding $k = 2$, the Expanded uncertainty U is

$$U = k \times u_c \qquad (6.42)$$

6.3 Summary

This chapter has introduced the measurement technologies of graphene-related materials by SEM and TEM, including electron microscope calibration reference materials, calibration methods, measurement methods and application cases of characteristic parameters of graphene sheet size, coverage, number of layers and lattice spacing by electron microscope. The technologies and methods involved have been carefully demonstrated and tested by the author. The data results are accurate and reliable. Not only can they provide theoretical basis for the research of graphene-related materials for researchers in research institutes, but they can also provide technical reference for the R&D of graphene-related materials and quality control in production enterprises. Based on the actual needs of the market and industry, the author will continue to do his share in the perfection of these measurement technologies and stay committed to applying these technologies and methods directly to graphene-based products in the market and industry. In this way, the maximum connection of technologies and methods with the market and industry can be realized, and the rapid and healthy development of China's graphene industry can be promoted.

References

[1] McCafffrey J P, Baribeau J M. A transmission electron microscope (TEM) calibration standard sample for all magnification, camera constant, and image/diffraction pattern rotation calibrations. *Microscopy Research and Technique*, 1995, 32(5): 449–454.

[2] Zhou J, Chen Z. A study of three-micron level grid graphics standard template used for SEM image magnification calibration. *Journal of Chinese Electron Microscopy Society*, 2005, 24(3): 185–191.

[3]   Qian J, Shi C, Tan H, *et al.* A study of calibration of SEM using one-dimensional grating specimen. *Journal of Metrology*, 2010, 31(4): 299–302.

[4]   Jiang B, Li Z M, Zhai H, *et al.* Influencing factors on the determination of Corundum particle parameters by SEM and image analysis. *Advanced Materials Research*, 2011, 412: 441–444.

[5]   Geim A K, Novoselov K S. The rise of graphene. *Nature Materials*, 2007, 6(3): 183–191.

[6]   Geim A K. Graphene: Status and prospects. *Science*, 2009, (5934): 1530–1534.

[7]   Meyer J C, Geim A K, Katsnelson M I, *et al.* The structure of suspendedgraphene Sheeets. *Nature*, 2007, 446(7131): 60–63.

[8]   Hernandez Y R, Gryson A, Blighe F M, *et al.* Comparison of carbon nanotubes and nanodisks as percolative fillers in electrically conductive composites. *Scripta Materialia*, 2008, 58(1): 69–72.

[9]   Park S, Ruofff R S. Chemical methods for the production of graphenes. *Nature Nanotechnology*, 2009, 4(4): 217–224.

[10]   Berger C, Song Z M, Li X B, *et al.* electronic confinement and coherence in patterned epitaxial graphne. *Science*, 2006, 312(5777): 1191–1196.

[11]   Emtsev K V, Bostwick A, Horn K, *et al.* Towards wafer-size graphene layers by Atmospheric pressure graphitization of silicon carbide. *Nature Materials*, 2009, 8(3): 203–207.

[12]   Yu Q K, Lian J, Siripglert S, *et al.* Graphene segregated on Ni surfaces and transferred to insulators. *Applied Physics Letters*, 2008, 93(11): 113103.

[13]   Zhao L Y, Levendorf M, Goncher S, *et al.* Local atomic and electronic structure of boron chemical doping in monolayergraphene. *Nano Letters*, 2013, 13(10): 4659–4665.

[14]   Lee Y, Bae S, Jang H, *et al.* Wafer-scale synthesis and transfer of graphene films. *Nano Letters*, 2010, 10(2): 490–493.

[15]   Nair R R, Blake P, Grigorenko A N, *et al.* Fine structure constant defines visual Transparency of graphene. *Science*, 2008, 320(5881): 1308.

[16]   Blake P, Brimicombe P D, Nair R R, *et al.* graphene-based liquid crystal device. *Nano Letters*, 2008, 8(6): 1704–1708.

[17]   Huang P Y, Ruiz-Vargas C S, van der Zande A M, *et al.* Grains and grain boundaries in single-layergraphene atomic patchwork quilts. *Nature*, 2011, 469(7330): 389–392.

[18]   Tsen A W, Brown L, Levendorf M P, *et al.* Tailoring electrical transport across grain boundaries in polycrystallinegraphene. *Science*, 2012, 336(6085): 1143–1146.

[19] Li X S, Zhu Y W, Cai W W, *et al.* Transfer of large-area graphene films for high performance transparent conductive electrodes. *Nano Letters*, 2009, 9(12): 4359–4363.

[20] Hao Y F, Bharathi M S, Wang L, *et al.* The role of surface oxygen in the growth of large single-crystalgraphene on copper. *Science*, 2013, 342(6159): 720–723.

[21] Wang H, Wang G Z, Bao P F, *et al.* Controllable synthesis of submillimeter single-crystal monolayergraphene domains on copper foils by suppressing nucleation. *Journal of the American Chemical Society*, 2012, 134(8): 3627–3630.

[22] Zhao Z J, Shan Z F, Zhang C K, *et al.* Study on the diffusion mechanism of graphene grown on copper pockets. *Small*, 2015, 11(12): 1418–1422.

Chapter 7

Measurement of Chemical Composition of Graphene Powder by XPS and ICP-MS

Renxiao Liu, Guanglu Ge, Peng Xu, and Guolan Tian

7.1 Overview

Graphene is a typical two-dimensional nanomaterial with excellent physical, mechanical and chemical properties. It has wide application prospects in various industrial fields, such as electronics,[1] photonics,[2] energy storage,[3] medical applications,[4] electrochemical sensing technology[5] and complex materials. In recent years, graphene is considered as a new type of strategic material by many countries. "Graphene" is defined in ISO/TS 800004-13:2017[6] as "single layer of carbon atoms with each atom bound to three neighbors in a honeycomb structure" and is also called graphene layer, monolayer graphene or single-layer graphene.

Two main types of graphene products have been used in industrial fields, namely graphene film prepared by deposition or epitaxial growth on a substrate and graphene powder. However, owing to the limitations of industrial techniques and processes, graphene products with perfect monolayer can't be mass produced till date. The main form of industrial graphene product in China is graphene powder, which consists of numerous graphene nanoplates with thickness of 1–3 nm and lateral dimensions of 100 nm–100 μm. More than 30 graphene industrial parks have been build up in China, and some factories have capabilities to mass-produce graphene powder. Graphene powder has practical applications in various industrial fields, such as electrode materials of lithium battery, conductive slurry, thermal conductive film and heavy-duty coating. Today, lots of efforts are being made to achieve large-scale industrial production of graphene for various applications.

Graphene powder can be synthesized using two routes: (1) top-down approach using natural or synthetic graphite as raw materials, mainly by oxidation–reduction, intercalation dissociation, liquid-phase exfoliation, mechanical exfoliation, etc. (2) bottom-up approach using small organic molecules as precursor, mainly by chemical vapor deposition (CVD), crystal epitaxy or chemical synthesis. There are distinct difference of physicochemical properties among graphene powders prepared through different manufacture techniques, hence, the test methods shall be set up to evaluate the key control characteristics (KCCs) of graphene powders, such as specific surface area, defect level, surface oxygen-containing functional group, carbon to oxygen ratio (C/O), oxygen content, metal impurities and ash etc. Standard measurement methods need to be developed with utmost priority to build agreement and recognition in the industrial chain and subsequently promote the high-quality development of graphene industry.

The measurement techniques for chemical composition analysis of graphene powder, such as auger electron spectroscopy (AES), X-ray photoelectron spectroscopy (XPS), electron energy loss spectroscopy (EELS), energy dispersive X-ray spectrum (EDS, EDX), thermal analysis technique (DTA, TG, DSC) and inductively coupled plasma-mass spectrometry (ICP-MS), should be selected based on the specific characteristics of graphene powder. This is because (1) graphene powder is easy to drift and disperse because of its low density, (2) the presence of numerous lattice defects, stacking and surface chemical groups in industrial graphene powder can result in complex microstructures and physicochemical properties, (3) although graphene is a typical carbon nanomaterial, the C content varies considerably among different types of graphene powders, such as graphene (G), graphene oxide (GO) and reduced graphene oxide (rGO), and (4) non-metallic elements, such as oxygen, sulfur, nitrogen, chlorine and silicon, and metallic impurities such as sodium, iron, copper, titanium, barium, tungsten, molybdenum, chromium and manganese are present in industrial graphene powders.

Based on the current trends in the graphene industrial field (with graphene powder as the main product form), various specifications, complex chemical compositions, etc. and considering the accuracy, reliability and generalizability of analytical techniques for the chemical composition analysis of graphene powder, this chapter mainly focusses on (1) XPS technique to measure C/O and (2) ICP-MS technique to determine concentration of metal impurities.

7.2   X-ray Photoelectron Spectroscopy

7.2.1   Introduction

An X-ray photoelectron spectrometer, which uses soft X-rays as the excitation light source, is highly sensitive to surface chemical properties, such as elemental composition, chemical state and distribution of surface elements. The XPS instrument consists of a vacuum system, sample delivery system, X-ray source, energy analysis system, detector and computer operating system. The working principle of XPS is shown schematically in Fig. 7.1.

When a test sample is irradiated with a beam of X-ray with sufficient energy (hv), such as Al K_α (1486.6 eV), the photons interact with the sample surface, and the total energy of the photons is transferred

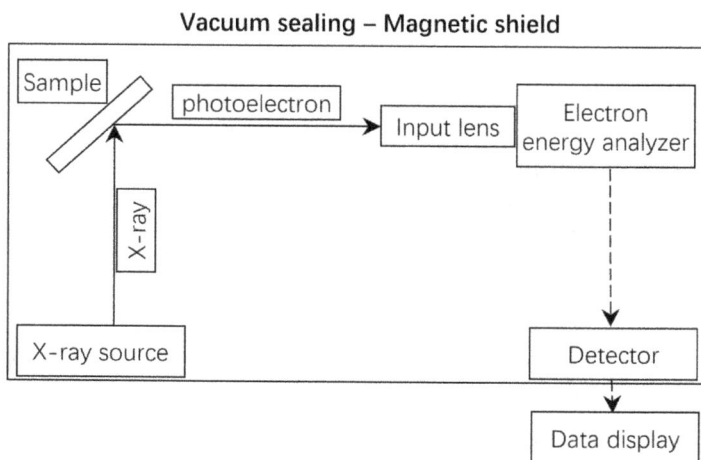

Fig. 7.1.   Schematic diagram of XPS working principle.

to the bound electrons in the atoms or molecules. The electrons of different energy levels are ionized with a specific probability resulting in the emission of photoelectrons with energies characteristic of the elements measured. The energy analyzer records the kinetic energies of the photoelectrons ejected from the surface of the test sample, and a graph of photoelectron count vs. binding energy is obtained. Thus, the elemental composition and chemical state of the material surface can be analyzed qualitatively and quantitatively using XPS. The XPS instrument has several advantages for the surface analysis of solid materials: (1) low test sample amount, (2) sample pre-treatment not required, (3) short testing time, (4) wide analytical range from atomic number 3 (lithium) to 92 (uranium), (5) chemical state can be determined, which helps in the analysis of compositions of materials and compounds, and (6) non-destructive test, suitable for analysis of organic materials and polymers.

XPS, also called electron spectroscopy for chemical analysis (ESCA), is an advanced technique for surface chemical analysis. The elemental composition of a solid surface can be analyzed quantitatively using XPS through integration of peak area by the relative sensitivity factor method; the detection sensitivity is about 0.1%.

7.2.2   XPS Instrument Calibration

The surface chemical properties, including elemental composition, chemical state and relative element content, can be obtained by analyzing the peak position, peak shape and intensity (peak height or peak area) of each characteristic peak in the XPS spectrum. To ensure the accuracy and reliability of the results, the XPS instrument is calibrated prior to the measurement. According to GB/T 25184-2010 *Verification Method for X-ray Photoelectron Spectrometers*, the calibration of the XPS instrument involves the calibrations of energy scale, intensity scale, selected area and spatial resolution of XPS imaging, charge neutralization, and sputtering rate of ion gun.

The calibration of the energy scale (in eV) includes the peak position (binding energy) and peak width of the characteristic peak in the XPS spectrum. The working parameters of the X-ray source and the energy spectrometer, including pass energy, deceleration ratio, and slit and lens parameters, are set according to the calibration requirements, and the binding energies of reference materials (RMs), namely

Cu $2p_{3/2}$ of copper foil and Au $4f_{7/2}$ of gold foil, are measured. The repeatability standard deviation of the peak binding energy and the linearity of the binding energy scale are available in GB/T 22571-2008 *Surface Chemical Analysis — Calibration of Energy Scales for X-ray Photoelectron Spectrometers*.

The calibration of the intensity scale includes the peak intensity, expressed as the peak height or peak area. The peak height is measured in counts or counts rate (counts per second), and the peak area is measured in counts · eV or counts · eV/s. Copper foil is used as the RM for the calibration of intensity scale. To measure the linearity of the intensity scale, the working parameters of each X-ray source and energy spectrometer, including pass energy, deceleration ratio, and slits and lens parameter, need to be selected. Since the calibration result is related to the working parameters, the linearity of the working parameters is evaluated using common analysis conditions. According to GB/T 21006-2007 *Surface Chemical Analysis — X-ray Photoelectron and Auger Electron Spectrometers — Linearity of Intensity Scale*, the linearity of the intensity scale should be measured using (a) the variable source flux method for the XPS instrument with X-ray flux set at 30 or more approximately equidistant increments and (b) the spectral ratio method for the XPS instrument with X-ray flux less than 30 predefined values. The repeatability and consistency of the intensity scale are calibrated in accordance with GB/T 28633-2012/ISO 24237:2005 *Surface Chemical Analysis — X-ray Photoelectron Spectroscopy — Repeatability and Constancy of Intensity Scale*. Copper foil is used as the RM, and the working parameters of the repeatability can be set as same as those of the linearity calibration to calculate peak area, peak intensity, intensity ratio and uncertainty. The consistency of the intensity scale is evaluated periodically, and the consistency is measured after the repeatability measurement.

The calibration of selected area and imaging spatial resolution is referred to ISO 15470 *Surface Chemical Analysis — Description of the Performance Parameters of X-ray Photoelectron Spectrum Selection Instrument*, ISO 18516 *Surface Chemical Analysis — Determination of Lateral Resolution and Sharpness based on Nanometer to Micron Scale in Beam-based Method*, and ISO/TR 19319 *Surface Chemical Analysis — Basic Method of Determining Lateral Resolution and Sharpness based on Beam Method*. At present, the optimal imaging spatial resolution of XPS is approximately $3\,\mu$m. The three methods used for the calibration of spatial resolution are as follows:

(1) **Method of characteristic zone:** When the test sample has a characteristic zone that is 30% less than the spatial resolution of the XPS instrument, the full-width at half-maximum of the characteristic curve of the photoelectron signal in this zone is defined as the spatial resolution.

(2) **Method of combination of straight edges of two different materials:** The test sample consists of two different materials, and the surfaces of the two materials form a common straight edge on the same plane. When measuring in the vertical direction to the common straight edge, the scanning line of the characteristic photoelectron intensity of one of the two materials is used to define the spatial resolution.

(3) **Edge method of single material:** The edge of the test sample of the material sheet covers half of a small hole with a diameter of approximately 1 mm and a depth not less than 5 mm. When measuring in the vertical direction to the edge, along the diameter of the small hole, the scanning line of the characteristic photoelectron intensity of the edge material is used to define the spatial resolution.

7.2.3   RMs for Calibration of XPS Instrument

Copper, gold and silver polycrystalline metal foils with purity higher than 99.99% are used as RMs for the calibration of energy scale of the XPS instrument. Prior to the measurement, the surface of the RM is cleaned by Ar sputtering till the peak heights of C 1s and O 1s are less than 2% that of the strongest metal peaks in the XPS spectrum, and the calibration is completed within one day. Notably, the area of Ar sputtering is larger than the measurement area, and the sputtering treatment should be appropriate; excessive sputtering may roughen the surfaces of the RMs and cause noticeable changes in the absolute emission intensity.

When calibrating the sputtering rate, the thickness of the RM is selected according to the characteristics of the test samples measured and the sputtering depth. At present, RMs of nanoscale film thickness, such as SiO_2 monolayer film formed by thermal oxidation on monocrystalline silicon, Ta_2O_5 monolayer or multi-layer film on polycrystalline Ta substrate, and AlAs/GaAs superlattice film on GaAs substrate, are used. Information on RMs used as measurement standards can be obtained online from the National Sharing Platform for RMs (http://www.ncrm.org.cn).

7.2.4   Notes for Measurement Using XPS Instrument

7.2.4.1   *Charge control and charge correction*

If the test sample is non-conductive, the surface area irradiated by the X-ray will exhibit charging effect. The degree of charging depends on several factors, including size, shape, installation and contact with the holder of the test sample, and requirement of charge control. The method used to perform charge control is selected according to the actual measurement conditions. In the presence of charging effect, the common correction method of binding energy measured will be described in detail to ensure its reproducibility and effectiveness. Neutralization is usually performed using a low-energy electron neutralization gun or neutralizer, especially for monochromatic XPS with severe charging effect. The methods for charge correction, such as differential charge, references of foreign contamination of hydrocarbons, gold deposition, inert gas injection, reference of internal standard and substrate reference, can be selected from the Appendix of GB/T 25185-2010 *Surface Chemical Analysis — X-ray Photoelectron Spectroscopy — Report on Methods for Charge Control and Charge Correction.*

7.2.4.2   *Ion gun*

The ion gun is used to clean the surface tested and perform sputter depth profiling. An Ar ion gun is commonly used. The calibration of the sputtering rate of the ion gun can affect the accurate depth measurement of the surface tested; hence, the sputtering parameters corresponding to each sputtering rate are provided when calibrating the sputtering rate of the ion gun (according to GB/T 20175-2006 *Surface Chemical Analysis — In-depth Analysis of Sputtering — Optimization Method of Layer Film System as Reference Material*).

7.3   Measurement of C/O of Graphene Powder

C and O are the major elements present in graphene powder, and C/O is a key parameter for characterization of graphene powder. For accurate measurement of C/O, an analysis method shall be established and a recognized measurement standard shall be developed. Graphene powder with higher C/O is classified as graphene and that with lower C/O is classified as GO.

XPS can be used to quantitatively measure the contents (atomic fractions) of C, O, N and S elements in graphene powder, and the C/O can be calculated using the atomic fractions of C and O elements. The following section describes the accurate and quantitative measurement of C/O of graphene powder using XPS.

7.3.1   Selection of Test Samples

The large-scale industrial production and application of graphene is underway, and lots of companies are involved in the production of graphene powders through different processes. Therefore, test samples are selected from the industrially produced graphene powders to promote the high-quality development of graphene industry. To ensure the reliability and universality of the measurement method and obtain a good statistical distribution, the selected test samples should be (i) typical representatives of different types of graphene powders and (ii) homogeneous and stable to exclude the uncertainty introduced by test samples. To measure the C/O of graphene powder using XPS, National Center for Nanoscience and Technology, China (NCNST), as the project leader, conducted Versailles Project on Advanced Materials and Standards (VAMAS) study in VAMAS/TWA41 (two-dimensional graphene-related materials). Three types of graphene powders covering a wide range of C/O values were prepared as test samples: mechanically exfoliated graphene, reduced GO and GO with nominal C/O values of 30, 5 and 2, respectively.

7.3.2   C/O Measurement Using XPS through VAMAS study

The round-robin test is commonly used to develop a universal measurement method and evaluate its performance. The VAMAS comparison study involving different international test laboratories is used to verify the reliability and reproducibility of the measurement method developed. Since the comparison protocol and test samples are the same, the VAMAS test results obtained from all the participating laboratories are analyzed statistically. Based on the test data obtained and problems encountered during the comparative study, the measurement protocol is improved and the sources of measurement uncertainty are evaluated to improve the measurement method as much as possible. The effectiveness of the VAMAS comparison study of measurement and analysis of KCCs of a new material can be ensured if (i) the measurement method is accurate and reliable and (ii) the test sample is homogeneous and stable.

The method for measurement of C/O of graphene powder using XPS is developed through VAMAS interlaboratory study. A feasible comparison protocol is developed based on the study of all the essential factors, including instrument calibration, sample preparation, measurement conditions and data processing.

(a) Measurement principle

The relative sensitivity factor method is adopted for quantitative analysis of C/O. The relative sensitivity factor of C 1s is taken as reference to obtain the relative sensitivity factors of other elements. If the relative sensitivity factors of two elements i and j in a solid sample are S_i and S_j, respectively, and the intensities of their characteristic peaks obtained using XPS are I_i and I_j (usually peak area used), respectively, then the ratio of their atomic fractions n_i/n_j is calculated using Equation (7.1):

$$\frac{n_i}{n_j} = \frac{I_i/S_i}{I_j/S_j} \qquad (7.1)$$

(b) Sample preparation

The graphene powder sample is dried at 80°C for 6 h in vacuum condition to remove the adsorbents on the surface, and the dried powder is pressed into tablets as test samples under clean conditions. The guidelines are provided in GB/T 28894-2012/ISO18117:2009 *Treatment of Samples Prior to Surface Chemical Analysis*. Sample contamination during preparation shall be avoided by using disposable gloves and masks. The test samples are quickly transferred into the preparation chamber of the XPS instrument, and when the vacuum reaches 5×10^{-8} mbar (5×10^{-6} Pa) or higher, the test sample is transferred to the analysis chamber to start measurement.

(c) Measurement procedure

(1) Calibrate the energy and intensity scales of the XPS instrument using Au, Ag and Cu polycrystalline foil RMs to meet the calibration requirements of the instrument.

(2) **Data collection:** First, perform a wide scan of the test sample at full energy range to obtain a survey spectrum with the maximum peak intensity $>2 \times 10^5$ count/s. Next, perform narrow scans of C 1s (278–296 eV) and O 1s (526–540 eV) at a scan step of 0.05 eV, with eight repetitive scans and the measurements repeated thrice.

(3) **Charge correction:** Perform charge correction by considering the lowest binding energy (248.8 eV) of C 1s peak as the standard.

(4) **Data analysis:** Integrate the areas of the peaks associated with individual elements. The integration range is 282–294 eV for C 1s and 529–539 eV for O 1s. The percentage contents (atomic fractions) of C and O elements are obtained using the relative sensitivity factor method, and then C/O is calculated.

(5) **Uncertainty evaluation:** Type B uncertainty introduced by the instrument can be neglected based on two aspects: (i) the XPS instrument has high precision and sensitivity and (ii) the XPS instruments of all the participating laboratories in VAMAS have already been calibrated. Thus, the main measurement uncertainty is Type A, introduced by the test samples and the measurement process. The measurement uncertainty can be analyzed directly using the measurement results of each participating laboratory of VAMAS.

7.3.3 Case Study

Taking the VAMAS comparison study as an example, the rGO powder sample was pressed into a 1-cm-diameter round tablet and placed on a silicon wafer for XPS measurements, as shown in Fig. 7.2. According to the measurement procedure in Section 7.3.2, the XPS spectra of C 1s and O 1s were obtained, as shown in Fig. 7.3.

The measurement results of the VAMAS comparison study obtained from 13 participating laboratories are shown in Table 7.1, wherein Table 7.1(a) is the percentage content (atomic fraction) of C element and Table 7.1(b) is the percentage content (atomic fraction) of O element; the number of test specimens or test points is different (based on the feedback by the participating laboratories).

First, the measurement data are analyzed statistically by classical statistics using the Grubbs method to check and remove any suspicious data. Second, the normality of all the data distributions is examined, and the average value of each group of data is considered as a single measured value to form a new group of measurement data. The suspicious values are again removed using the Grubbs method. Third, equal accuracy is determined using the Corcoran method. Finally, all the measurement data are analyzed using PauTa criterion. In the absence of any suspicious data, the average percentage contents and standard deviations of atomic fractions of C and O elements are calculated using all the measurement data, and then the C/O of the rGO test sample is obtained, as shown in Table 7.2.

Fig. 7.2. Digital photo of rGO test sample.

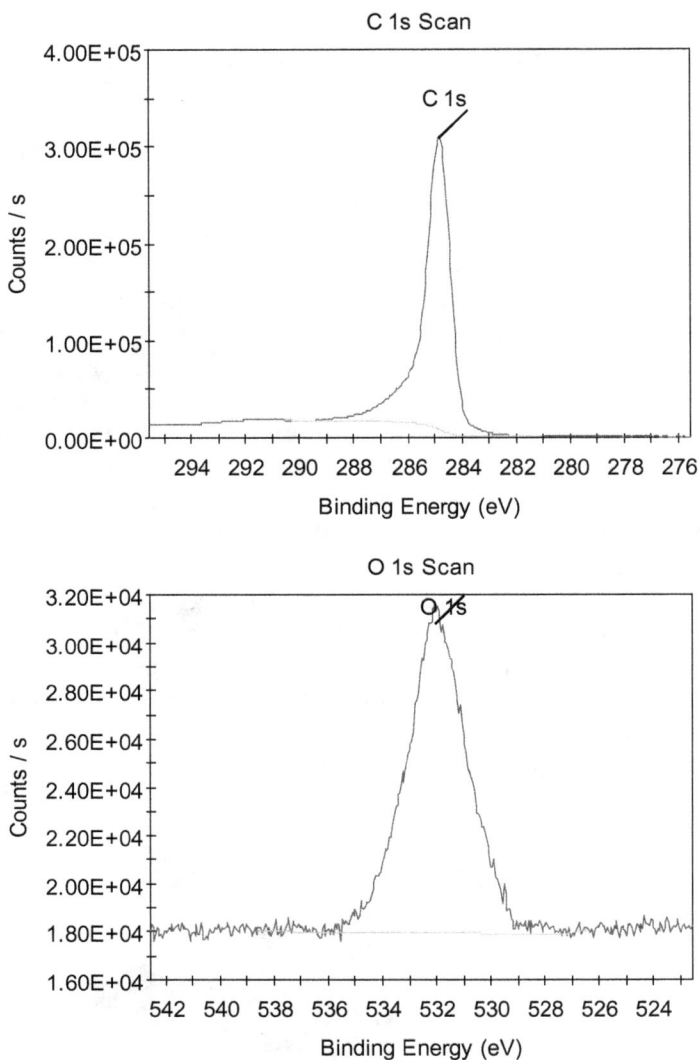

Fig. 7.3.   XPS spectra of C 1s and O 1s of rGO test sample.

The average percentage content of O element and C/O obtained from all the participating laboratories of VAMAS are analyzed by robust statistics using the median and standard interquartile range (IQR) values as measures of data centralization and dispersion.

Table 7.1(a). Percentage content (atomic fraction) of C element from VAMAS study.

No.	L1	L2	L3	L4	L5	L6	L7	L8	L9	L10	L11	L12	L13
1	84.75	84.7	85.53	85.02	83.62	82.46	86.31	84.64	83.29	84.25	81.85	83.59	85.05
2	85.54	84.75	85.59	84.98	82.79	83.38	85.16	84.11	83.74	84.35	82.38	83.7	84.99
3	85.37	84.51	84.99	84.38	83.16	82.89	85.52	84.82	82.22	84.11	82.17	83.59	85.00
4	85.33	84.48	85.21	85.23	83.64	82.53	84.49	—	81.99	84.13	82.41	83.72	85.06
5	84.96	84.85	84.76	83.7	83.53	82.44	85.2	—	82.88	83.77	82.38	83.57	84.93
6	85.17	84.36	85.22	84.5	83.41	82.04	85.01	—	83.78	84.14	82.28	83.85	84.78
7	85.44	84.45	84.69	85.5	83.13	82.62	85.74	—	81.75	84.2	82.38	83.67	85.01
8	84.89	84.45	84.86	85.16	84.5	83.03	86.03	—	83.25	84.25	82.11	83.83	84.92
9	84.79	84.76	84.91	85.09	83.46	82.6	85.56	—	81.27	84.03	82.17	83.73	84.88
10	84.63	84.51	85.84	—	83.66	—	86.18	—	82.07	84.19	82.59	83.72	85.00
11	84.6	84.64	84.86	—	—	—	—	—	82.93	—	82.19	—	84.97
12	84.73	84.66	85.01	—	—	—	—	—	—	—	—	—	—
Ave.	85.02	84.59	85.12	84.84	83.49	82.67	85.52	84.52	82.56	84.14	82.26	83.70	84.97
SD	0.34	0.15	0.36	0.55	0.45	0.39	0.57	0.37	0.87	0.16	0.19	0.10	0.08

Table 7.1(b). Percentage content (atomic fraction) of O element from VAMAS study.

No.	L1	L2	L3	L4	L5	L6	L7	L8	L9	L10	L11	L12	L13
1	14.49	14.86	14.17	13.99	15.88	16.29	13.26	14.92	15.28	17.68	15.33	14.45	14.49
2	14.43	14.98	14.20	14.17	16.63	15.52	13.75	15.46	15.18	17.35	15.25	14.54	14.43
3	14.59	15.08	14.75	14.80	16.34	15.74	13.48	14.64	15.42	17.33	15.41	14.54	14.59
4	14.55	15.26	14.61	13.90	15.81	16.20	14.06	—	15.38	17.3	15.25	14.42	14.55
5	14.40	14.92	15.03	15.10	15.94	16.40	13.53	—	15.66	17.38	15.32	14.57	14.40
6	14.27	15.20	14.50	14.05	16.00	16.79	13.7	—	15.43	17.35	15.21	14.70	14.27
7	13.83	15.07	15.05	13.67	16.11	16.47	13.4	—	15.31	17.35	15.36	14.53	13.83
8	14.08	15.20	14.89	13.95	14.95	15.57	13.28	—	15.22	17.54	15.23	14.60	14.08
9	14.31	14.91	14.90	13.98	15.78	16.24	13.77	—	15.42	17.28	15.34	14.61	14.31
10	13.99	15.16	13.97	—	15.72	—	13.11	—	15.28	17.17	15.30	14.55	13.99
11	13.91	15.04	14.92	—	—	—	—	—	—	17.37	—	14.60	13.91
12	13.75	15.01	14.74	—	—	—	—	—	—	17.56	—	14.47	13.75
Ave.	14.22	15.06	14.64	14.18	15.92	16.14	13.53	15.01	15.36	17.39	15.30	14.55	14.22
SD	0.29	0.13	0.36	0.46	0.44	0.43	0.29	0.42	0.14	0.14	0.06	0.08	0.29

Table 7.2.   The average percentage content of C and O elements, and C/O of rGO test sample.

C element		O element		C/O	
Average percentage content	84.24	Average percentage content	15.12	Average value	5.57
Standard deviation	1.06	Standard deviation	1.07		

The Z-score is calculated using Equation (7.2), and the degree of data concentration is represented by the robust coefficient of variation (CV):

$$Z = \frac{\text{Measured value} - \text{median value}}{\text{Standard IQR}} \tag{7.2}$$

The median value for a set of measurement data arranged from small to large is the middle number for an odd set and the average value of two middle numbers for an even set:

$$\text{Standard IQR} = \text{IQR} \times 0.7431 = (Q_3 - Q_1) \times 0.7431$$

where Q_3 is the upper quartile and Q_1 is the lower quartile. Robust CV = Standard IQR/Median value × 100%.

Taking the Z-score as the judgment criteria, $|Z| \leq 2$ is satisfactory, $2 < |Z| < 3$ is suspicious and $|Z| \geq 3$ is rogue.

The measurement results and robust statistics of C/O and O contents of the rGO test samples are shown in Table 7.3.

The measurement results of classical statistics and robust statistics are similar, which indicates that the laboratories participating in the VAMAS comparison study have high and equivalent measurement capabilities. Furthermore, the results of robust statistics show that all the Z-scores of the measurement results obtained from all the participating laboratories except Lab 11 (owing to the limited statistical properties of few data points) are less than 2. Therefore, the measurement results of the VAMAS comparison study are accurate and credible, the test sample is sufficiently homogeneous, and the comparison protocol is reliable.

Table 7.3.   The robust statistics results of C/O and O content
of rGO test sample.

Laboratory	C/O	Z-score	O content	Z-score
L1	5.98	0.82	14.22	−0.81
L2	5.62	0.00	15.06	0.01
L3	5.82	0.45	14.64	−0.40
L4	5.99	0.84	14.18	−0.85
L5	5.25	−0.84	15.92	0.85
L6	5.13	−1.11	16.14	1.07
L7	6.29	1.52	13.53	−1.49
L8	5.63	0.02	15.01	−0.04
L9	4.86	−1.73	17.02	1.93
L10	5.47	−0.34	15.36	0.30
L11	4.73	−2.02	17.39	2.29
L12	5.47	−0.34	15.30	0.25
L13	5.84	0.50	14.56	−0.49
Median	5.62	—	15.06	—
Minimum	4.73		13.53	
Maximum	6.29		17.39	
Range	1.56		3.86	
Standardized ICR	0.44		1.02	
Robust CV	7.83%		6.77%	

7.4   Inductively Coupled Plasma-mass Spectrometry

Inductively coupled plasma-mass spectrometry (ICP-MS) is a type
of trace analytical technique with high sensitivity used for quantitative measurement of inorganic elements, which can simultaneously
detect multiple elements. The basic principle of ICP-MS is as follows: the nebulized test sample is introduced into a plasma torch by
a carrier gas, and the test sample can be dissolved, atomized and
ionized under the high-temperature action of plasma. The ions are
separated in the mass spectrometer according to their mass–charge
ratio. For an element with a certain mass–charge ratio, the signal
intensity (represented as counts per second) is proportional to the
concentration of the element in the test sample solution; hence, the
contents of all elements in the test sample solution can be measured
quantitatively.

The quantitative measurement of ICP-MS is based on the peak
area of each element measured, and the mass fraction X_i of each

metal impurity can be calculated using Equation (7.3):

$$X_i = \frac{V_S I_d (C_{S,i} - C_{0,i})}{M_S} \tag{7.3}$$

where V_S is the final volume of the test sample solution (L), I_d is the dilution factor of the test sample solution, including all the pretreatment steps, $C_{S,i}$ is the mass concentration of any metal impurity i in the test sample solution (mg/L), $C_{0,i}$ is the mass concentration of any metal impurity i in the blank solution (mg/L) and M_S is the mass of the test sample of graphene powder (g).

To ensure the accuracy and reliability of the measurement results, the standard recovery is evaluated. The standard recovery R_i, of a single element i is calculated using Equation (7.4):

$$R_i = \frac{C_{S',i} - C_{S,i}}{C_i} \tag{7.4}$$

where $C_{S',i}$ is the concentration of element i in the spiked solution of the test sample (mg/L), $C_{S,i}$ is the concentration of element i in the original test sample solution (mg/L) and C_i is the concentration of element i in the standard solution added (mg/L).

At present, no calibration regulation for ICP-MS instrument has been developed; however, different kinds of stock solutions and calibration standards are available for status evaluation of the ICP-MS instrument. Prior to measurement, the operation parameters and indexes of the ICP-MS instrument are calibrated using a tuning solution to ensure that all the indexes, including sensitivity, oxide, double charge and resolution, meet the measurement requirements. Subsequently, the measurements are performed according to the operation procedure.

7.5 Determination of Metal Impurities Contained in Graphene Powder

Owing to the differences in raw materials, processes and facilities, a variety of metal impurities (over 20 species) are present in industrial graphene powder. The species and content of the metal impurities vary with the type of graphene powder, manufacturer and process. The application performance of the graphene powder in certain industrial fields is significantly affected by the metal impurities; hence, an accurate method for measurement of metal impurities is necessary. Similar to the C/O measurement using XPS, a quantitative measurement method for metal impurities contained in graphene powder using ICP-MS was developed using the VAMAS comparison study.

Three kinds of industrial graphene powders were selected as test samples for the VAMAS study: (i) rGO, same as the XPS test sample, (ii) GO, same as the XPS test sample, and (iii) few-layer graphene produced by small molecules growing.

7.5.1   Preparation of Test Samples

To get accurate and reliable measurement results, the preparation method of test samples shall be considered as well as the capability and status of the measurement instrument, especially for the measurement of metal impurities contained in graphene powder with complex elemental composition and obvious matrix effect. Therefore, prior to ICP-MS measurements, the graphene powder sample is pretreated. In view of the various species and wide content distribution of metal impurities, the wet chemical digestion method combined with microwave digestion is recommended for test sample pre-treatment.

Strong acids, such as ultra-pure concentrated nitric acid (HNO_3) and hydrofluoric acid (HF) (for samples with high silicon oxide content), are used as the digestion solvent. The test samples of graphene powder are digested in a microwave digester under constant high temperature ($>200°C$) and high pressure (set according to instrument capability). The final digested solution should be clear and transparent, which generally requires 1–3 cycles of digestion depending on the test sample. Four to six parallel test samples are recommended, including 1–2 spiked test samples to determine the standard recovery. The standard recoveries of metal impurities with high content, such as Fe, Cr, Ni, Mn, Cu, Zn, Na, Mg, Al, K and Ca, are measured. The standard recovery shall be generally 90–110%, but the graphene powder exhibits significant matrix effect and much lower homogeneity than liquid samples, therefore, a standard recovery of 80–120% is reasonable and acceptable for test samples of graphene powder.

7.5.2   ICP-MS Measurement of Graphene Powder

The measurement procedure includes two steps: (i) perform a quick survey scan for all the inorganic elements in the test sample solution to identify the species and estimate the concentration level of metal impurities qualitatively and semi-quantitatively and (ii) classify all the metal impurities and perform quantitative measurement using ICP-MS.

(1) According to the species and concentration level of the metal impurities, select several standard stock solutions and dilute them with 2% HNO_3 solution to obtain a series of suitable concentrations and establish a standard curve.
(2) Several metal impurities with equivalent concentration levels can be measured simultaneously based on the standard curves of multiple elements. The slope, intercept and correlation coefficient of the standard curve are calculated by the linear regression method. The correlation coefficient should be ≥ 0.99.
(3) Set the measurement method and measure the concentration of metal impurities in the test sample solutions and in contrast solution.

7.5.3   Case Study of Metal Impurities in Contained Graphene Powder

Taking VAMAS study using ICP-MS as an example, 20 milligrams of industrial rGO powder was weighed and placed in the microwave digestion vessel. Eight milliliters of concentrated HNO_3 was added dropwise, and the closed vessels were assembled into the microwave digester. Six parallel test specimen were prepared. The microwave digestion process was performed by programmed heating, and the maximum temperature was maintained at 195°C for 30 min. A clear solution was obtained after 3 cycles of digestion, which was evaporated using a hot plate to approximately 0.5 mL left. This solution was diluted with 2% HNO_3 solution in a 10-mL volumetric flask.

The suitable internal standard elements (Table 7.4) and standard stock solutions (Table 7.5) are selected according to the survey scan results.

The ICP-MS VAMAS study is in progress, and Table 7.6 shows one of the measurement results obtained. Up to 23 species of metal

Table 7.4.   Internal standard elements for ICP-MS measurement of rGO test sample.

No.	Internal standard element	Mass number	Theoretical value (μg/kg)	Recovery of internal standard (%)	Elements with equivalent mass number to internal standard
1	Ge	74	~20	~95.49	Na Mg K Ca Fe Al Cr Mn Ni Cu Zn V Co Ga As Rb Sr
2	Rh	103	~20	~98.36	Ag B Ti Zr Mo Nb
3	Re	187	~20	~97.12	Ba Pb Hf Au
4	In	115	~20	~98.36	La, Ce, Pd, Sn, Sb,
5	Tb	159	~20	~96.80	Ta W

Table 7.5.   Stocks standard solutions for ICP-MS measurement of rGO test sample.

RM number	Concentration (mg/kg)	Elements	Solvent	Uncertainty
GBW(E) 082429	10.0	Ag, Al, As, Ba, Be, Bi, Ca, Cd, Co, Cr, Cs, Cu, Fe, Ga, In, K, Li, Mg, Mn, Na, Ni, Pb, Rb, Se, Sr, Tl, U, V, Zn	5% HNO_3	3% ($k = 2$)
GBW(E) 082428	10.0	Ce, Dy, Er, Eu, Gd, Ho, La, Lu, Nd, Pr, Sc, Sm, Tb, Th, Tm, Y Yb	5% HNO_3	3% ($k = 2$)
GBW(E) 082430	10.0	Au, Hf, Hg, Ir, Pd, Pt, Rh, Ru, Sb, Sn, Te	1% HNO_3 plus 10% HCl	3% ($k = 2$)
GBW(E) 082431	10.0	B, Ge, Mo, Nb, P, Re, S, Si, Ta, Ti, W, Zr	0.9% HNO_3 plus 0.9% HF	3% ($k = 2$)

impurities are detected in the rGO test sample. The data histogram of Table 7.6 is shown in Fig. 7.4, which reveals the differences in the contents of the various metal impurities.

Table 7.6.   ICP-MS measurement results of rGO test sample as case study.

No.	Element	Mass number	Content, mg/kg						Average
			6 parallel test specimen						
1	Na	23	5,375	5,478	5,014	5,941	6,210	6,183	5,700
2	Mg	24	126.06	141.72	137.34	136.92	155.6	127.24	137.48
3	Al	27	5.16	21.81	5.37	8.70	15.56	12.68	11.55
4	K	39	168.02	179.28	169.16	173.00	170.12	163.98	170.59
5	Ca	43	339.85	410.00	228.35	258.70	326.90	254.60	303.07
6	Ti	47	6.75	11.18	8.36	7.57	11.75	8.13	8.96
7	V	51	0.23	0.28	0.25	0.22	0.22	0.22	0.24
8	Cr	52	13.91	15.55	14.05	12.89	13.55	13.05	13.83
9	Mn	55	8.28	8.79	8.14	8.06	8.11	8.34	8.29
10	Fe	57	68.68	91.50	74.76	65.98	95.62	72.52	78.18
11	Co	59	0.08	0.10	0.07	0.07	0.07	0.08	0.08
12	Ni	60	3.67	3.71	4.38	3.01	3.72	3.08	3.60
13	Cu	63	0.22	0.79	0.35	0.26	0.49	0.35	0.41
14	Zn	66	3.71	6.84	2.58	1.83	6.02	2.09	3.85
15	Rb	85	0.36	0.43	0.36	0.36	0.37	0.37	0.38
16	Sr	88	2.42	2.94	2.34	2.53	2.46	2.74	2.57
17	Zr	90	6.97	7.22	6.58	6.94	6.86	6.93	6.92
18	Mo	98	4.86	5.32	5.06	5.16	5.22	5.08	5.12
19	Pd	106	0.20	0.18	0.16	0.18	0.16	0.18	0.18
20	Sn	118	0.11	0.12	0.09	0.09	0.11	0.09	0.10
21	Sb	121	0.11	0.21	0.11	0.22	0.14	0.20	0.17
22	Ba	138	0.90	1.41	0.97	0.97	1.21	1.19	1.11
23	Hf	180	0.27	0.19	0.16	0.17	0.17	0.17	0.19

The reliability of the measurement results is evaluated by the standard recovery. Figure 7.5 shows the standard recoveries of some of the metal impurities contained in the rGO test sample. The standard recoveries of most of the metals except Ni are within the acceptable range of 80–120%. The abnormal value for Ni presumably arises from the Ni cone of the ICP-MS instrument.

7.6   Establishment of Measurement Standard

The measurement standards for the KCCs of graphene material can promote related industrial applications. The measurement results of the VAMAS study can confirm the reliability and applicability of the comparison protocol and the universality and representativeness of

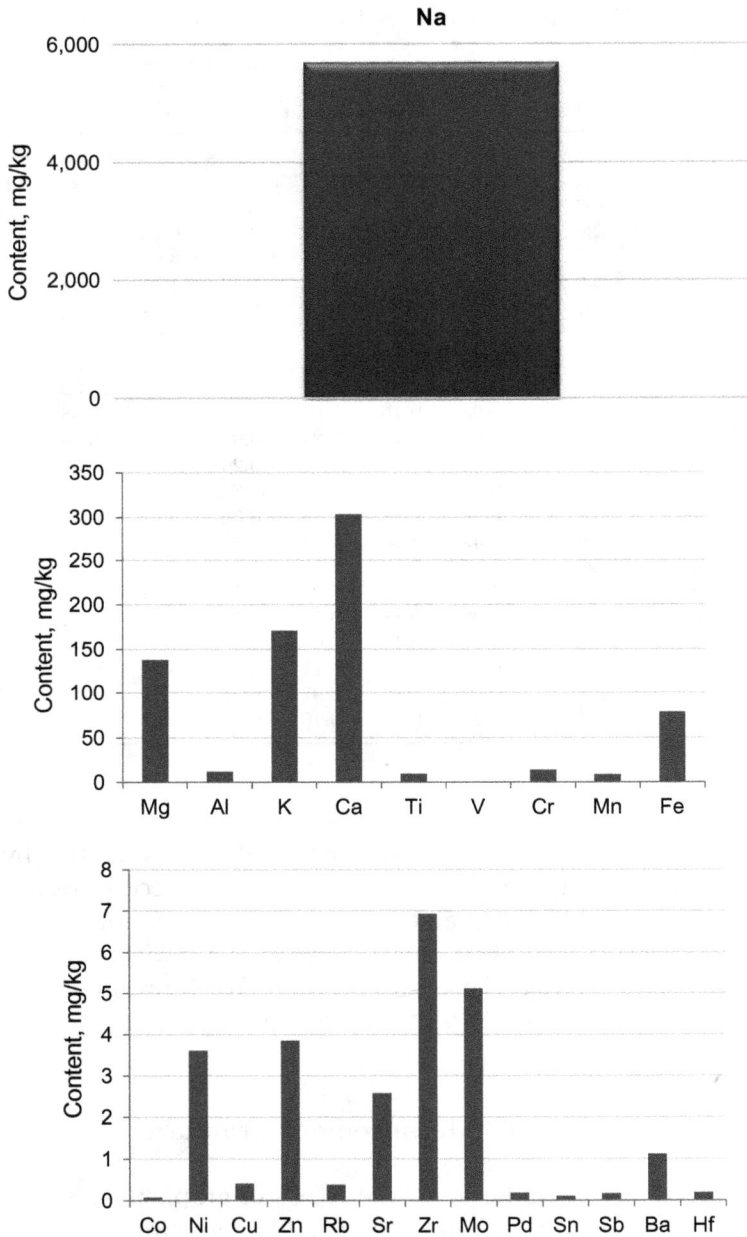

Fig. 7.4.   The species and contents of metal impurities contained in rGO test sample.

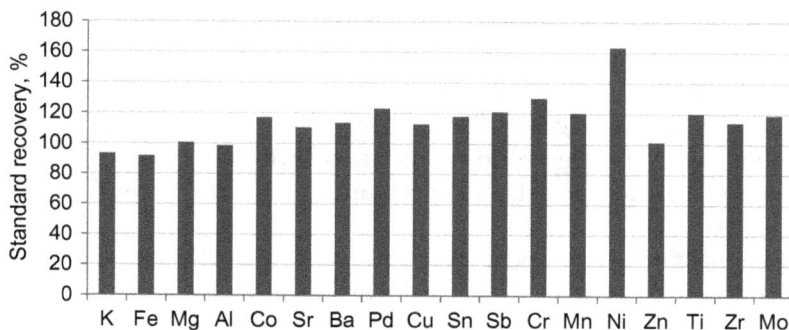

Fig. 7.5. Standard recovery of some metal impurities contained in rGO test sample.

Table 7.7. Two international standards of graphene powder developed in IEC/TC113.

Standard-1	IEC/TS 62607-6-20:2022
	Nanomanufacturing — Key control characteristics — Part 6-20: Graphene-based material — Metallic impurity content: Inductively coupled plasma mass spectrometry
Standard-2	IEC/TS 62607-6-21:2022
	Nanomanufacturing — Key control characteristics — Part 6-21: Graphene-based material — Elemental composition, C/O ratio: X-ray photoelectron spectroscopy

the test samples of industrial graphene powders. In addition, the laboratories participating in the VAMAS study have high measurement capabilities; the measurement results exhibit good comparability and also possess international mutual recognition. Based on the VAMAS study, two international standards have been developed and published in IEC/TC113 *International Electrotechnical Commission Nanotechnology for electrotechnical products and systems*, as shown in Table 7.7.

7.7 Summary

As a typical two-dimensional nanomaterial, graphene with excellent electronic and thermal properties has been explored by many countries in recent years. Efforts for mass production of graphene film are currently in progress, and graphene powder has realized industrial preparation and application on a large scale in several industrial fields, such as conductive slurry, heavy-duty coating, high abrasion tire and new type electrode materials of lithium battery. Graphene materials exhibit excellent mechanical, electrical and thermal properties.

The properties of industrial graphene are inferior to those of theoretical graphene with perfect carbon atom monolayer. This is attributed to (1) the numerous defects and stacking faults within the micro-structure of industrial graphene produced using current techniques, which significantly decrease the performance of mechanical, electronic and thermal applications, and (2) impurities introduced into industrial graphene materials during manufacture, such as chemical reagents of acid, alkali, surfactants and stabilizers used in top-down routes and organic and inorganic impurities arising from catalysis and processing conditions of bottom-up routes. Hence, the standard measurement methods for the KCCs described in this chapter are relevant for industrial applications of graphene materials. In addition, other standard measurement methods of chemical composition, such as surface oxygen-containing functional groups, anion content and sulfur content, are being developed.

References

[1] Weiss N O, *et al.* Graphene: An emerging electronic material. *Advanced Materials*, 2012, 24(43): 5782–5825.
[2] Bonaccorso F, *et al.* Graphene photonics and optoelectronics. *Nature Photonics*, 2010, 4(9): 611–622.
[3] Pumera M. Graphene-based nanomaterials for energy storage. *Energy and Environmental Science*, 2011, 4(3): 668–674.
[4] Krishna K V, *et al.* Graphene-based nanomaterials for nanobiotechnology and biomedical applications. *Nanomedicine*, 2013, 8(10): 1669–1688.
[5] Ambrosi A, *et al.* Electrochemistry of graphene and related materials. *Chemical Reviews*, 2014, 114(14): 7150–7188.
[6] ISO/TS 80004-13:2017 Nanotechnologies — Vocabulary — Part 13: Graphene and related two-dimensional (2D) materials.

Index

versailles project on advanced
materials and standards (VAMAS),
29, 32, 50, 146, 223–225, 229,
232–233, 235

W

World Trade Organization (WTO),
41

X

X-ray diffraction (XRD), 24, 57–58,
100, 103–104, 106–107, 197
X-ray diffractometer (XRD), 31, 43,
55
X-ray photoelectron spectroscopy
(XPS), 55, 216–218, 220, 223–224,
232